# PRAISE FOR
# FORGING AHEAD

"*Forging Ahead* is an inspiring testament to how strong values can guide a business for generations, touching the lives of not only its employees but countless others. The dedication and faithfulness of the Nix family to values-driven leadership make this an inspiring read for entrepreneurs and anyone who needs a tangible example of how staying true to one's faith in business can lead to immense success in both business and life."

—**TOM MONAGHAN**, Founder of Domino's Pizza,
Ave Maria University and School of Law, Legatus

"A delightfully written book about the little business that could."

—**JULIA SCHEERES**, *New York Times* bestselling author

"The Nix family's journey in *Forging Ahead* illustrates how legacy is built over generations. Matthew Nix's vision to grow NIX Companies while preserving its values offers a powerful example for business leaders. *Forging Ahead* is a valuable resource for those wanting to create a lasting culture while balancing tradition with ambition."

—**SCOTT FARMER**, chairman of the board, Cintas

"Becoming a fifth-generation family enterprise requires a lot of work in the family and in the business. *Forging Ahead* shares the journey of the Nix family with the world and provides other business families with insights on how to build and maintain a generational legacy."

**—ISABEL BOTERO**, PhD, associate professor and chair in family entrepreneurship at the University of Louisville

"Angie Klink and Matthew Nix have masterfully woven together the rich history of the Nix family and their business legacy. Klink skillfully crafts the lens to gaze into the portals of family-business life. With a passion for serving God and people, the Nix family has blended spirituality, passion to succeed, and the drive to foster success in their employees and communities. Reading Klink's testament to the Nix family will awaken your entrepreneurial emotions to bring your life to an enhanced state of grace."

**—IRA BOOTS**, former chairman and CEO, Berry Plastics; chairman of the board, Milacron

"*Forging Ahead* is a captivating narrative that beautifully intertwines the rich history of Nix Companies with Matthew Nix's dynamic leadership. It is an inspiring and deeply personal journey through evolution and transformation. Angie Klink's remarkable storytelling brings to life the challenges, triumphs, and unwavering values that have propelled this family business to incredible heights. *Forging Ahead* is not just about business growth; it is about the heart and soul of a company that values its people above all. It is a must-read for anyone seeking inspiration from a true American success story."

—**TERESA ROCHE**, PhD, chief human resources officer, City of Fort Collins

"*Forging Ahead* is a story about faith and family, purpose and passion—all beautifully told by an extraordinary storyteller. This inspiring, highly readable account offers valuable lessons about leadership, humility, and what it takes to achieve the American dream."

—**TERI RIZVI**, founder, Erma Bombeck Writers' Workshop, author of *One Heart with Courage: Essays and Stories*

*Forging Ahead:*
*How Five Generations of Small-Town Values Collided with Big Ambitions to Spark One of America's Fastest-Growing Companies*

by Angie Klink

© Copyright 2024 Angie Klink

ISBN 979-8-88824-675-7

All rights reserved. No part of this publication may be reproduced, stored in a retrieval system, or transmitted in any form or by any means—electronic, mechanical, photocopy, recording, or any other—except for brief quotations in printed reviews, without the prior written permission of the author.

Cover photo: Carl A. Nix Sr., second generation, works at the anvil at Nix Welding. (Photo from *Evansville Courier*, 1977.)

Published by

3705 Shore Drive
Virginia Beach, VA 23455
800-435-4811
www.koehlerbooks.com

# FORGING AHEAD

How Five Generations of Small-Town Values
Collided with Big Ambitions to Spark One of
America's Fastest-Growing Companies

## ANGIE KLINK

VIRGINIA BEACH
CAPE CHARLES

To Steve.
Because we, too, know love, dedication, and pride
in marriage and family business.

"If you like apple pie, you're gonna keep eatin' apple pie. I just liked weldin', and so I just kept weldin'.

"We all know that life is just an attitude. You can make it whatever you want to. We were always fair. They'd say, 'Bid on this job.' I'd say, 'No, I'd rather *not* bid on it. I'll just do what it takes. That way I don't get tattooed, and you don't get tattooed. We'll keep each other fair, and we'll always be friends that way.'"

—Carl A. "Sonny" Nix
Third-generation owner, Nix Welding

"We create jobs that have meaning, purpose, and dignity. We want people to love where they work and feel productive. And that is what matters."

—Matthew Nix
Fifth-generation CEO, Nix Companies

Except where otherwise noted, all Matthew Nix quotes in this book are taken from *Be Big, Act Small*, an unpublished manuscript written by Matthew Nix.

## WHAT IS NIX?

Nix Companies Inc. is a family holding company focused on service-oriented businesses, real estate, and other supporting services.

Nix Industrial is the legacy family business dating back to 1902 that evolved from a blacksmith shop to a welding shop and today is a custom manufacturer and industrial repair company focused on metal fabrication, machining and gears, and coatings and finishing.

## CONTENTS

*Preface* ............................................................. xvii
*Introduction* ........................................................ xx
*The Magic of Welding* ............................................... xxiii

Chapter 1: Alloy of Faith and Family ................................. 1
Chapter 2: Dad Needs My Help at the Shop ............................. 6

**Part 1: Family Foundation** ........................................ 11
Chapter 3: First Generation .......................................... 13
Chapter 4: Second Generation ......................................... 20
Chapter 5: Third Generation .......................................... 26
Chapter 6: Fourth Generation ......................................... 33
Chapter 7: Fifth Generation .......................................... 48

**Part 2: Fifth Generation Finds the Way** ........................... 55
Chapter 8: Paid to Figure It Out ..................................... 57
Chapter 9: Maternal Combustion ....................................... 65
Chapter 10: You Just Know What You Know .............................. 75
Chapter 11: A 100-Year-Old Start-up .................................. 80
Chapter 12: Rite of Passage .......................................... 85
Chapter 13: Yacht in a Cornfield ..................................... 93
Chapter 14: Boat-Building Blues ...................................... 100
Chapter 15: Doubling the Business Overnight .......................... 103
Chapter 16: First Salesperson ........................................ 112
Chapter 17: The Paint Shop ........................................... 118
Chapter 18: An Opportunity to Build Something ........................ 124
Chapter 19: With Me or Against Me? ................................... 129

Chapter 20: Learning to Lead ..................................................... 136
Chapter 21: Rock Bottom .......................................................... 145

**Part 3: Precision-Shaping the Future** ........................................... 153
Chapter 22: Search for Meaning .................................................. 155
Chapter 23: Setting the Plan in Motion ..................................... 163
Chapter 24: The Value of Underutilized Potential ....................... 170
Chapter 25: Return to Poseyville ................................................ 181
Chapter 26: It's About the People ............................................... 192
Chapter 27: Company Culture: The Vibe Inside ........................ 200
Chapter 28: Pandemic Shifts ...................................................... 206
Chapter 29: Irons in the Fire ...................................................... 216
Chapter 30: Keeping the Family Feel ......................................... 227

*Epilogue* ...................................................................................... 239
*Acknowledgments* ....................................................................... 245
*Appendix* .................................................................................... 247
*Endnotes* ..................................................................................... 253

# PREFACE

I received a phone call from writer and friend Amy Abbott in the winter of 2023, asking me if I would be interested in writing a book about a company in Southern Indiana. She had no idea what the company was. She simply knew they needed a writer.

Amy, from Newburgh, Indiana, had been contacted by Tad Dickel of T. A. Dickel Group, an acquaintance who had worked with NIX on leadership enhancement. Tad asked Amy if she would be interested in writing a book about this "mystery company" and his client, the "fifth-generation owner of a family business outside of Evansville," whom he admired greatly.

I knew Amy from the Erma Bombeck Writers' Workshop at the University of Dayton. Named after the late syndicated newspaper humor columnist Erma Bombeck, this glorious workshop is where writers gather to fortify their writing and one another amid much laughter. Since the workshop, we have stayed connected through social media in a collaborative, supportive thread fostered by our shared experience and the aura of Erma. That nourishing thread reminds me of the people-centered NIX company culture you will read about in this book.

Writing a book about a company and its family legacy was not in Amy's wheelhouse, but she kindly thought of me, as the subject sounded similar to that of *Kirby's Way: How Kirby and Caroline Risk Built Their Company on Kitchen-Table Values*, my book about the founders of the Kirby Risk Corporation. Amy gave me a call, and we chatted about what little she knew of the project. Frankly, not knowing what company wanted a book made me hesitant. Many people approach me with book ideas, and most will not come to fruition.

So, when I received an email from Tad asking to connect me with the company, I delayed replying. Fortunately, he wrote again about a week later. I decided to respond, and I'm very happy I did.

I soon discovered that Matthew Nix, CEO of Nix Companies, was all in. He was dedicated to creating a book that would tell the inspiring, fascinating story of his family's business, which began as a blacksmith shop, then became a longtime welding shop until Matthew entered the picture. His drive for growth escalated his family business into a regional company on its way to becoming a national name in metal fabrication and diversified commercial and industrial metal services while headquartered in tiny, quaint Poseyville, Indiana, since 1902.

I can relate to the roller coaster ride of a family business legacy because my husband, Steve, and I once owned Indiana's oldest pharmacy, founded in 1829 in Lafayette, Indiana. Wells-Yeager-Best Pharmacy had been owned by my husband's father, John Klink, who passed away suddenly in 1980. While not a pharmacist, Steve, at age twenty (about the age Matthew Nix was when he started going gung-ho with NIX), was attending Purdue University at the time, and even though he had always said he was not going to work in the family business, he took over. During the following twenty-six years, he grew the business into a small, successful local chain.

Wells-Yeager-Best Pharmacy was a part of our family. It had a soul and a personality that sat nightly with us at our family dinner table and was a living, breathing part of the Greater Lafayette community. While we sold the business in 2007, the wonderful relationships we established with our customers, staff, and hometown linger as a legacy for which we are grateful.

Thank you to my four early beta readers: Michelle Kreinbrook, community benefit and outreach director, North Central Health Services, Inc./River Bend Hospital; David Beering, owner of Intelligent Design; Dennis Carson, economic development director for the City of Lafayette;

and Kitty Campbell, executive director of Leadership Lafayette.

And a special thank-you goes out to Amy Abbott, Tad Dickel, Matthew Nix, and the entire Nix Companies family for the opportunity to tell the NIX story. For me to write about NIX, a family business headquartered in small-town Indiana, was an honor and a joy. A relatable kinship.

—Angie Klink, November 2023

# INTRODUCTION
by Matthew Nix

## WHY A BOOK?

I agreed to share the NIX story in writing for two reasons. One, it has been said that you don't truly know a person until you know where they come from. As our organization continues to grow and we add new team members, this book will be a valuable resource for them to get a sense of the heart and soul of our organization. Moreover, I hope it inspires our team members to write the best chapters of their own stories, whether or not they end up in the pages of a book. Two, I felt the book might be relatable to others outside the company on their own career walk, particularly those involved in a family business or any leader of a small or midmarket enterprise who aspires to grow. As you will read in the subsequent pages, I have been deeply influenced by several such stories. If this book can help one person the way other books have helped me, it is a worthwhile endeavor.

• • •

## ABOUT OUR AUTHOR AND THOUGHTS ON THE BOOK

I was delighted to be connected with Angle Klink through Tad Dickel (T. A. Dickel Consulting Group) and thankful for her genuine interest and excitement in telling our story. Her professional background as a historian and author, along with her firsthand experience as part of a family business, made her a perfect fit. She did an eloquent job of sharing our story in an entertaining, relatable, and authentic way.

Angie worked many hours conducting interviews, reviewing hundreds of pages filled with recorded accounts, and researching through newspaper articles dating back to the early 1900s. She spent a couple of days visiting Poseyville, touring our operations, and meeting with team members. I believed it was important for her to really "feel" what our organization was about, and she ably captured her impressions on paper for the reader.

I knew Angie had really captured the story when Dad (a natural critic and not one to contact me to chitchat) called me to say he had read the first nine chapters from the deer stand while out hunting and the remaining chapters the next day. "She did a good job," he said. "I haven't read that much in my whole life!"

Just as I feel called to the work I do, Angie is certainly living out her calling. Whether it's a schoolteacher, a nurse, a welder, or a writer, it's always a pleasure to be around someone who is truly "on mission." We are thankful she chose to share her time and talents with us.

While our story accelerated in recent years, the strong foundation was built more than 100 years and four generations before us. It's an honor to carry on that legacy and not something that I take lightly. While much of this book highlights my vision and ambition, that in and of itself was not nearly enough to build what we have today. In fact, you can't build much of anything on vision and ambition alone. The most important thing I have done in my time leading our company is to build our team. Nothing else comes close.

The essence of our story lies in our mission statement, four simple words that have become our mantra: "Be Big, Act Small." You will learn much more about this concept throughout the pages of this book. My brother, Adam, created the phrase, doodling it during a group brainstorming session. It's fitting he developed it, because he and others on our team balance my natural inclinations, my drive, with their more methodical, careful oversight. Together we are the yin and the yang, if you will, of NIX. Without both styles and the foundation wrought by the previous four generations of the Nix family, there would be no growth story to tell.

There have been many businesses that grew like a rocket ship but quickly evaporated—or businesses that treated everyone around them as a means toward their own end. On the other hand, there have been many humble, values-based organizations that had much to contribute to the world but left it largely untapped and wasted. What a shame! At NIX, we aspire to be big and do great things but also to act small and do it the right way.

What I appreciate most about the storyline Angie has presented is that it represents much more than corporate success; it embodies family, faith, stewardship, and the enduring legacy of American enterprise. All aspects of life that I hold dear.

# THE MAGIC OF WELDING

Most of us go through our days not even thinking about how welding, of all things, affects so many aspects of our lives. Todd Bridigum, author of the book *How to Weld*, deems today's metal-joining technology as "extraordinary."

"What if all the welds in the world came apart at one time and welders weren't available to repair them?" Bridigum asks. "Society as we know it would fall apart, or at least electricity would go away."[1]

Those are thought-provoking words. Welding keeps America running.

"Power plants are held together by welds and maintained by pipefitters, ironworkers, boilermakers, and machinists," he continues. "All of us rely on electricity and on those skilled professionals for our standard of living. . . . I believe metal fabrication is the single largest influence on how the world looks today; from the cities to the country and even into space."[2]

Forge welding, the first metal-joining process, was developed during the Iron Age. Those who transformed rock into iron tools, known today as blacksmiths, were thought to possess magical powers.[3] Nix Companies originated as a blacksmith shop owned by Charles Henry Nix, a man with magical powers.

# CHAPTER 1

# Alloy of Faith and Family

St. Francis Xavier Catholic Church stands sentinel at the end of Main Street in Poseyville, Indiana—watching, protecting, inspiring. The redbrick church with its steeple topped by a gold cross sits at the "T" where St. Francis Street meets Main. Facing the church, one must choose a direction, perhaps a metaphor for those who grow up and decide whether to stay or leave Poseyville: turn north to head out of town toward Interstate 64 or turn south to meander slow country roads.

Poseyville is a Mayberry, Andy-and-Opie-Taylor sort of burg where the name Nix, the family and the company, has been a part of the community's fabric since the turn of the twentieth century. All five generations of the Nix family have been members of St. Francis Xavier Catholic Church. The family's Catholic identity and shared values helped shape the tenets of the Nix family's time-honored company.

It is a nostalgic kind of town, where family comes first—where honest, well-paying jobs can be found away from the bureaucracy of big business, with its crush of work time stealing from personal time. In Poseyville, population less than 1,000, you know your neighbor. Your kids play baseball with your coworkers' kids. Surrounded by wide-open wheat and corn fields, it's a place where you can breathe. And much of this idyllic good life is made possible through the company founded in a wooden building at the center of town in 1902: a blacksmith shop owned by Charles Henry Nix, a German Catholic immigrant now buried with his family in the St. Francis Cemetery.

St. Francis Xavier Catholic Church, where generations of the Nix family have been members, watches over Main Street in Poseyville, Indiana.

That blacksmith shop where Charles worked his entire life forging metal into farm implements and horseshoes with a hammer and anvil has grown to become a multimillion-dollar Midwest business with multiple locations and operating divisions. Nix Companies, formed in 2016, is the parent company of Nix Industrial, a custom manufacturing and industrial repair enterprise now run by the fifth generation of the Nix family, Matthew Nix.

When he was only nineteen, Matthew began reinventing his great-great-grandfather's business, which was handed down as a welding and machine shop to Matthew's grandfather and father. Matthew and his team took the 101-year-old family-owned company and in less than twenty years grew the revenue more than 100 times. And they did it in modest Poseyville, where all five generations of Nix men have lived, worked, and raised their families. How did he do it? *Why* did he do it?

Matthew has a longing, an insatiable drive to expand beyond his predecessors'—his ancestors'—wildest dreams. Dreams his grandfather and father never imagined, as they were content each day with helping local farmers, most of them longtime friends, by quickly and expertly repairing their broken implements to get them back in the fields, then locking the shop door in the evening and strolling home down the sidewalks of Poseyville to a family dinner with nary a thought of wanting more.

Matthew's father, Bill, a thick, round-shouldered man with curly hair and a dimple on his chin, is matter-of-fact. If he gives you a hard time with a few expletives thrown in, it means he is fond of you. Matthew claims that his personality is distinct from his father's—that they have "philosophical differences" and are "nothing alike." However, that is not completely true. It was Bill who instilled Matthew's impeccable work ethic. Neither Matthew nor his father are afraid of hard work. It's just that his father, for the most part, sees progress as more headaches and challenges. Matthew sees progress as opportunity. With all his successes in growing the company, Matthew still wrestles with this dichotomy as he strives to make the Nix name a national brand. He wonders why he is not content with the way things are, as is. Good enough. Sometimes Matthew envies his grandfather and father. Their lives were simpler.

Through prayer, motivational books, and self-reflection, Matthew has come to the realization that God made him this way—with an ambition to create an exceptional national company that treats its workers well, with compassion and genuine caring for their futures and their families. Matthew's calling is to change people's lives for the better through opportunities offered in an empathetic and supportive, yet competitive, workplace.

"And frankly, to not embrace this wouldn't be right," Matthew explained. "I'm a big believer in 'To whom much is given, much is expected.' That's easier said than done. But we have an opportunity and a platform to do a lot of good. We create jobs for people that have meaning, purpose, and dignity. We want people to love where they work and feel productive. And that is what matters."

How did Matthew and his team increase the revenue of a simple welding and machine shop a hundredfold? First, by creating a clear and compelling vision of the future. Then, by developing a world-class corporate culture forged from five generations of small-town values. The culture of Nix Companies is built on esteeming its people, celebrating its workers, and helping them realize their individual visions for a good life by fostering their dreams of satisfying jobs, children, a house, car, boat, a cabin at the lake—whatever makes them happy—while making sure their position at NIX is a good fit for them and the company. When people are happy, company success and well-calculated expansion follow.

This book reveals how Matthew and his like-minded leadership team—which today includes his wife, Lindsey, and his brother, Adam—learned through trial and error. By failing, facing their fears, and coming out the other side to focus not on herky-jerky growth but on the company's humanity. As St. Francis Church keeps watch over Main Street.

FORGING AHEAD

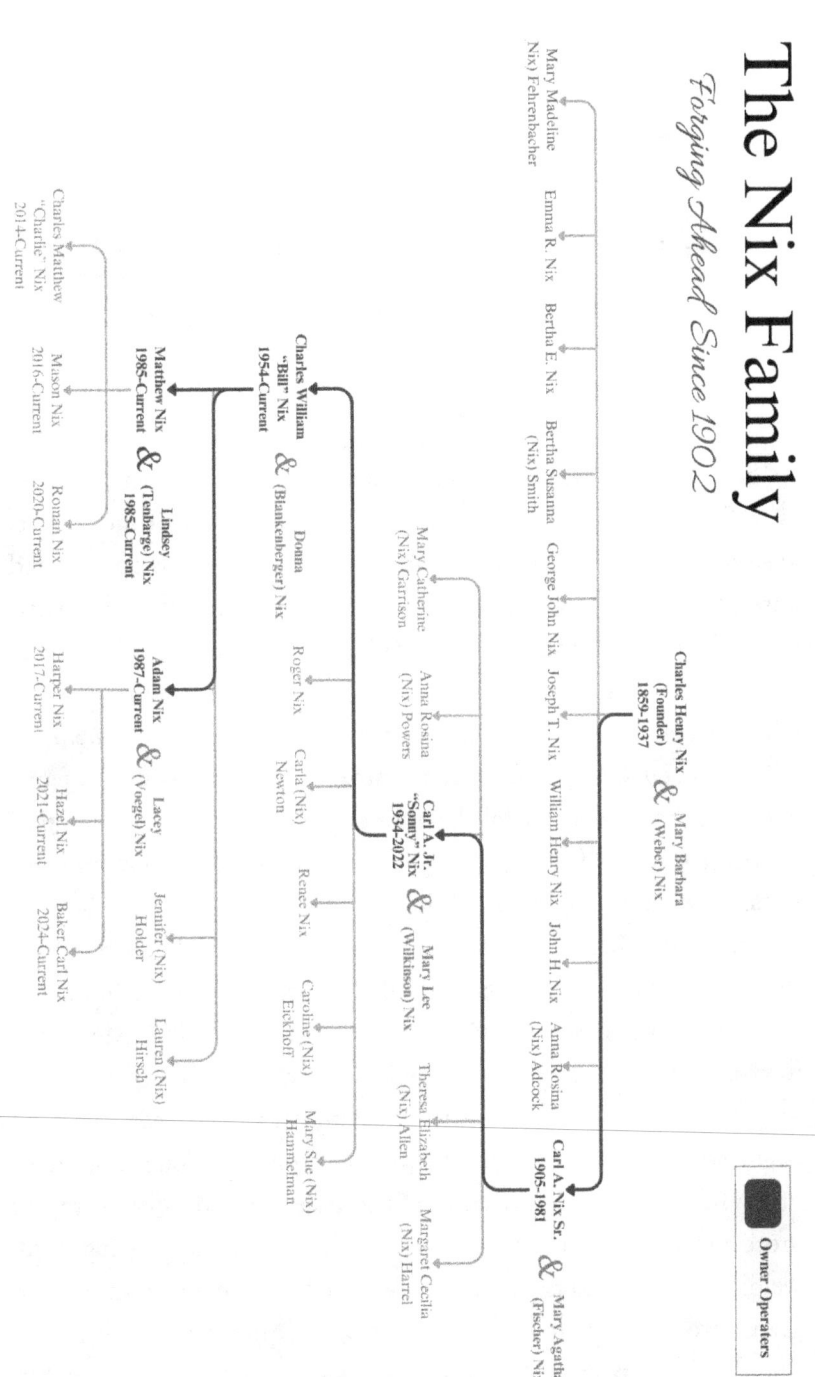

## CHAPTER 2

# Dad Needs My Help at the Shop

The daycare phone rang on a humid summer day in 1993. Carolyn, a family friend who cared for Poseyville children in her home, answered. After a quick conversation, she hung up and approached dark-haired Matthew, age eight, who was playing in a Nerf gun battle on the backyard swing set. "Your dad just called, and he will be here in a minute to pick you up. He needs your help at the shop."

Matthew's heart lifted. Thoughts swam in his head. *Dad needs my help? With what?* He envisioned his father in his dark-blue work pants and shirt dusted with metal shavings and grime, wearing his hood as he welded, sparks flying. How could Matthew, a little boy, help his father at what his family called the "welding shop"?

Bill picked up Matthew at daycare, and they headed to the pole barn that sat next to the cement-block building Matthew's great-grandfather and grandfather had built with their bare hands in 1957. Measuring forty-eight by forty-eight feet, the pole barn was where Matthew's father worked alongside Grandpa Sonny and Aunt Caroline. Matthew would begin working there in about ten years and cut his teeth on the business.

When little Matthew entered the pole barn, the first thing to greet him was the familiar shop smell: an acrid mix of cutting fluid, welding smoke, and hydraulic oil, with a hint of manure and rotten grain thrown in, compliments of the farm equipment in for repair. The shop's temperature hit him like a wall of heat from Hades, but he was a child, and that day he did not care that the shop had no air-conditioning for Indiana summer scorchers. All Matthew could think was *Dad needs my help.*

Charles William "Bill" Nix and Matthew Nix, circa early 1990s.

In 2023, Bill, now sixty-nine, recounted that day in his Southern Indiana lilt: "I had a flatbed trailer in there, and I had a hole cut in the top of the floor. You had to get your hand through that hole to tighten up the bolts on the kingpin. My hand was too big. I couldn't get my hand up in there to get the nuts on. So I went down to daycare and got Matthew. I knew he could get his hand up in there, and he put them nuts on there. There were probably eight or ten bolts that we had to tighten up, and it had to be done right because that was the kingpin. That's what hooks to the truck, and that's what pulls the whole trailer."[4]

Matthew remembered, "Dad was replacing dozens of rivets in a flatbed semitrailer with new bolts. He had drilled out the rivets and was in the process of installing the bolts, but they were in a very tight spot, so he needed a set of small hands to hold the nuts on the backside of the bolts. I was just the person for the job. I helped him with the project for the rest of the day and went home that evening with my clothes filthy and my chest swelled with pride."

The fact that Bill thought young Matthew could help him with a job spoke volumes for his confidence in his son, for Bill has always been a perfectionist. With his machinist background, he was always a stickler for details in his shop, taking pride in repairing a piece of equipment so that it was better and stronger than when new. In Bill's view, to fix something and have it break again in the same place was the ultimate sin. When Bill repaired a piece of equipment, it never returned.

Matthew was the oldest of four children. Two boys came along to Bill and Donna—a registered nurse—and then two girls. Matthew's earliest memories of the welding shop are the pit stops he and his brother, Adam, three years his junior, made there after the school day ended at North Elementary, located up the hill from the shop. They would make a beeline for the refrigerator to grab a candy bar and a Dr Pepper before continuing another block past their grandfather's huge vegetable garden next to the shop and on to their grandparents' house, where Grandma Mary Lee Nix cared for them until their father got off work. The three locations—school, shop, and their grandparents' house—were within blocks of each other and remain there today.

"Dad and Grandpa kept the refrigerator in the shop stocked," Matthew said. "There was a little metal tray inside the fridge where customers could leave fifty cents for a drink and a candy bar. It was the honor system, of course. After getting our sugar fix and visiting with Grandpa for a bit, we would be on our way."

A few years later, as a teenager, Matthew would begin working at the shop, sweeping the floor, emptying scrap buckets, and cleaning metal chips out of the machines. He reminisced:

> I can remember like yesterday riding to work with Dad on Saturday mornings. We would stop at T-mart [the gas station] and get a bacon or sausage biscuit and a Yoo-hoo. The shop had two old furnaces in the block building, and the newer pole barn didn't have any heat. I can remember getting there in the

mornings during the winter and turning up the thermostat so the heaters would kick on. You could see your breath in the shop.

At 9:30, we took a ten-minute break to have a Dr Pepper, some peanut butter crackers, a candy bar, or, during hunting season, deer salami and crackers. We chatted about hunting, sports, or whatever was good at the time and then got right back to work. Now I cherish the moments we spent together . . . ten minutes at a time.

Those snippets of relaxed break time offered a deep well of personal and workplace teachings subtly handed down by the third and fourth generations of the Nix family: Grandpa Sonny, Bill, and Aunt Caroline—Matthew's godmother, who did the books and worked on the shop floor.

As they chatted and slugged Dr Peppers, the spirits of the first and second generations lingered within the foundation.

Caroline Nix Eickhoff, Matthew Nix, Charles William "Bill" Nix, and Carl A. "Sonny" Nix Jr. in the "welding shop," circa 1990s.

Left, cement-block building built in the 1950s by Carl A. Nix Sr. and Carl A. "Sonny" Nix Jr. Right, metal pole barn added in the 1980s by Charles William "Bill" Nix.

# PART 1

## Family Foundation

# CHAPTER 3

# First Generation

## CHARLES HENRY NIX: "SUCCESS CROWNED HIS PERSISTENCY"

The founder of what would become Nix Companies started out his career working at Steinway Piano Company in New York City. Charles Henry Nix, the great-great-grandfather of current NIX CEO Matthew Nix, was fourteen when he emigrated from Germany in 1873. Born in Ramsthal in 1859, he came to the United States to escape the Kaiser's Army. Charles was a big fellow and destined for the Elite Guard when he turned eighteen.[5] One could say he was a "draft dodger," or one could say he was brave and smart to hightail it out of Germany. Eventually his parents and thirteen siblings also arrived in America.

When Charles arrived and started working at Steinway, he could not speak English, so he studied signboards hanging along the streets as he rode the horse-drawn streetcar to and from the Queens piano company in "Steinway Village."[6] In a few months, he had learned the language of his adopted country. What he did for Steinway remains a mystery.

When he was about eighteen, he moved to Ingraham, Illinois, where his parents and most of his siblings lived, and worked on a farm owned by the family of Mary "Barbara" Weber. While there, Charles became seriously ill and ended up in the family's clinic. Barbara, a dark-haired young woman with thin lips and plump cheeks, acted as his nurse during the long illness. A relationship developed, and they married.

The couple settled down in Ingraham, and Charles, strong yet graceful, farmed for several years before opening a blacksmith shop there in 1887. At six foot four and 250 pounds, with dark hair and a thick Teddy Roosevelt mustache, Charles was formidable as he forged glowing-hot iron into submission with blows from a hammer, sweat rolling down his back.

Charles offered general blacksmithing, wagon making, and repair at reasonable prices and sold machinery by Deering Harvester Company, which later merged with McCormick Harvesting Machine Company to create International Harvester. He outsold his competitors. An Ingraham newspaper explained, "He sells more implements simply because he buys in larger quantities and for cash, and, in consequence, can offer inducements not possible by other dealers."

For reasons unknown, Charles and Barbara moved the seventy miles from Ingraham to Poseyville in 1902. Traveling by horse and buggy, Charles brought all of his and Barbara's worldly possessions, along with $10,000, which he carried in a valise. His obituary noted, "He put [the valise] on a chair by his side as he ate his dinner in a Main Street restaurant." Charles owned his house in Ingraham free and clear, so the cash likely came from the sale of his home and business in Illinois.[7]

He purchased one of the four blacksmith shops in town, a small one-story wooden structure on Church Street, a dirt thoroughfare. He and Barbara lived nearby, one block south of downtown Poseyville, in the vicinity of the current post office.

Back then, a forge, anvil, hammer, tongs, and a vise were all the equipment a blacksmith needed. Four years later, Charles converted the first building into a shop to store and repair carriages and built a larger cement-block building to house his growing business, which included International Harvester farm "machines." After work, Charles often went to the pool hall to play pinochle, then, to his children's delight, brought home a bucket of root beer.

Original blacksmith shop owned by Charles Henry Nix
in downtown Poseyville, circa 1920s.

In the first twenty years of their marriage, Charles and Barbara had nine children, including three nonbiological children the couple took in and raised. Their son Carl A., Matthew's great-grandfather, was born in 1905. When he was about eight—the same age Matthew would be in 1993 when his father picked him up from daycare to help at the welding shop—Carl started working in the shop: shoeing horses, repairing and building wagon wheels, and shaping plowshares (the main cutting blade of a plow).[8] He would take over his father's blacksmith shop in 1924 when he was only nineteen.

Besides running the blacksmith shop, Charles also grew cantaloupe and watermelon on a farm northwest of Poseyville. The melons were crated and loaded onto trucks or a train refrigerator car and taken to market as far north as Chicago. Future generations of the Nix family would hold summertime jobs on melon farms. Still today, Southern Indiana is known for its large, delicious melons. The *New York Times*–acclaimed book *The Growing Season* by Sarah Frey chronicles another compelling family business story with its roots tied to Posey County melons.

Charles Henry Nix, first-generation blacksmith.

In September 1908, Williams Jennings Bryan, the Democratic candidate from Illinois running for president of the United States for the third and last time (he lost to William McKinley), arrived in Poseyville by train and was greeted by a crowd of 10,000 people "wild with enthusiasm," according to the *Evansville Courier*. He spoke from a pavilion erected in Fletchall's grove.⁹ Today, Nix Companies is located at 129 West Fletchall Avenue.

The headline read, "Was Treated to 60-Pound Watermelon, the Produce of a Posey County Farm." The story elaborated, "Before leaving Poseyville, seven large specimens of the famous yield of watermelons of Posey County were presented to Mr. Bryan by Charles Nix of Poseyville. A sixty-pounder was luscious eating for the party on the way to Evansville."¹⁰

In 1912, Charles, age fifty-two, was elected a member of the state legislature on the Democratic ticket for Vanderburgh and Posey Counties. By this time, his five sons were working at his business. An *Evansville Courier* article from 1913 bore the headline "Representative Nix Brings Up Sons to Work." The story states:

> Charles Nix of this place, who so ably represented Posey County in the lower house of the Indiana legislature during the 1913 session, has five interesting boys and like their father they have been brought up on work. . . . In fact, the Nix shop is one of the busiest places in Poseyville, which is a live and progressive town.
>
> The youngest son, Carl, can do anything from running a gasoline engine to shoeing a horse. . . . "We all work at my shop," said Mr. Nix. "And there is never an idle moment."¹¹

But work ethic was far from his only legacy. Charles's obituary spoke of his altruistic ways: "His most outstanding characteristic was his sympathy for the poor and needy. He was known throughout the community for his many charities which included many Catholic institutions and extended to persons many of whom never knew their benefactor. Few appeals for help were ever denied by Mr. Nix, who gave generously to many a worthy cause regardless of creed and belief."

A few years before he was elected to the state legislature, Charles showed his bent for leadership and perseverance on behalf of his church when he began corresponding with Andrew Carnegie, the nineteenth-century industrialist who helped build the American steel industry. This poor young man from Scotland immigrated to America and became one of the richest men in the United States. Carnegie is perhaps best known for giving money to build more than 2,500 public libraries, including one in Poseyville in 1904.[12]

In the summer of 1906, Charles wrote Carnegie regarding a possible donation toward purchasing a $1,000 pipe organ for Poseyville's St. Francis Xavier Catholic Church. Charles was a man of music. He and his sons all played the violin, and he owned a German model made in 1617, valued back then at $5,000. In December, Charles finally received word from Carnegie that he would pay half the cost; the church would pay the other half.

A 1906 story in *The Boonville Standard* states, "This is indeed gratifying news to our citizens, and the church is to be congratulated on its good fortune, and Mr. Nix is deserving much praise for his interest and success in the matter. He wrote two or three letters before any reply was received but never gave up hope and continued his efforts until success crowned his persistency."[13]

Because of his abiding love for St. Francis, Charles Nix received the honor of having his name inscribed on one of the church's stained glass windows. Still in use today, the St. Francis tracker-action pipe organ was built by the German Edmund Giesecke, Evansville's first organ builder.[14] It was refurbished in 2002 and is one of the few original organs in the Evansville diocese.[15]

Coming full circle, in 2023, Charles's great-great-grandsons Matthew and Adam and their wives, Lindsey and Lacey, contributed an endowment to St. Francis. They earmarked the donation for facility upkeep with a special note that the family prefers the money be kept aside for the preservation of the organ.

Charles died on November 3, 1936, and was taken to the funeral parlor. His children were called home. However, at the funeral home,

Charles revived. All of his children were en route and could not be reached, so when they arrived in Poseyville, they learned that their father had *not* died. Seven months later, on July 4, 1937, Charles passed away—as the family story goes—"for good."[16]

Even in death, Charles Nix, the first of five generations of Nix Welding CEOs, was persistent.

## CHAPTER 4

# Second Generation

### CARL A. NIX SR.: PIONEERING, SELF-TAUGHT WELDER

His hands became calloused and dark from decades of working with metal and building coal fires in his forge. Gloves only got in the way. A short man, Carl A. Nix Sr., the youngest son of Charles and Barbara Nix, began his career as a 200-pound blacksmith who built a fire in his forge by whittling wood shavings, lighting them, and then scooping coal upon the fire. By 1977, when he was seventy-two, he was using gasoline and coal to create a quick blaze. "The coals have to get red, just like the coals on a barbecue grill," Carl explained.[17]

Carl took over the blacksmith shop when he was nineteen. One hundred years later, his great-grandson Matthew would start working full-time at that same age, going on to influence the growth of what became Nix Companies.

Carl was the youngest of five boys, with three sisters interspersed between them. His two oldest brothers, George and Joe, originally took over the business from their father, but in 1924, in order to pursue other ventures, they transitioned the helm to Carl. George then opened Nix's Hardware Store, which was much like a general store, carrying everything from toys to china, in downtown Poseyville. It closed in 1979. Joe sold farm machinery in a building that became the Nix Chevrolet garage.[18]

Charles Henry Nix with his sons, including his youngest, Carl A. Nix Sr.,
who took over the blacksmith shop when he was nineteen.
(Photo from *Evansville Courier*, 1913.)

For many decades, Carl muscled the same kind of work over an open fire that his father had. In describing how to shoe an inhospitable horse, he once explained, "We had stocks where you'd mount the horse between two bars and lift him off the floor where he had no foothold. We had equipment to hold his foot. That was for what I call a mean horse, one that would kick, bite, paw, and swing his head to try to hit you. We used to get $2.50 for four new shoes nailed on the feet."[19] One day Carl was shoeing a cantankerous horse that rammed itself into the pointed end of the anvil and, sadly, was impaled.

Eventually, Carl saw farm equipment evolve from horse and plow to small steel-wheel tractors and then to the modern equipment of today. He opened the first International Harvester dealership in town, and his business progressed with the times, outlasting twenty-seven other "smithy" operations.[20]

Carl invented the power threader, although he did not tell the world. Had Carl patented it or produced the threader in quantity, the welding shop likely would have gone in a different direction. Matthew Nix explained, "Back then you had to put threads on bolts or pipe by turning a thread die by hand. My great-grandfather Carl used materials around the shop to make a fixture and vise to hold the part to be threaded and an old electric motor to do the turning. Many years later, they bought an actual pipe-threading machine that was an improved version of what Carl had developed, essentially the exact same concept."

In 1941, Carl, age thirty-six, purchased the company's first welding machine. This state-of-the-art equipment was a turning point for the metal industry, the Nix family, and their Poseyville business. Without Carl's forward-thinking and pivotal purchase, the third generation likely would not have joined the business, and Nix Companies would not exist today.

Interior of blacksmith/welding shop, circa 1940.

When Carl purchased the welding machine, the first of its kind in the area, he did not know how to weld. Wearing his soot-stained coveralls and his thick leather work boots while holding a lit cigarette between his lips, he taught himself by trial and error. In 1949, Carl discontinued shoeing horses and worked solely with acetylene and arc welding.[21] Today, this rusted machine with a center lever that regulates the "amperes" from low to high is preserved and displayed with other historical NIX ephemera in the company's headquarters.

Interior of original blacksmith/welding shop. Carl A. Nix Sr. purchased the company's first welding machine in 1941.

A 1977 *Evansville Courier* story quotes Carl as saying, "I worked in the shop and helped my dad from the time I was old enough to tighten a bolt. . . . But blacksmithing is almost out of the picture due to modern machinery. Instead of using a forge and anvil, you just throw it on the welding bench.

"It wasn't a bad life," Carl adds. "If I was able, I'd still enjoy coming down here [the welding shop] and doing blacksmith work rather than go on a picnic."[22]

Carl A. Nix Sr. working a manual metal shear, now on display in the lobby of Nix Industrial. (Photo from *Evansville Courier*, 1977.)

The muscles in Carl's arms and hands told the story of his decades of swinging a hammer and the tough labor he took in stride as a kind and gentle man. He loved to tell jokes and tease people around the potbellied stove in the shop. Bill, his grandson and Matthew's father, recalled, "He would say, 'Ya wanna hear a dirty joke?' And I'd say, 'Yea!' And he'd say, 'Little boy fell in a mud puddle.'"

Carl married Mary Agatha Fischer from nearby Mount Vernon. They had four daughters and a son, Carl Jr., also known as Sonny, who would one day take over the welding shop and pass it down to his son, Bill.

Carl was granted a deferment from serving in World War II because his business was considered essential. Like his father, Carl was a faithful Catholic attending St. Francis Xavier Catholic Church, going to Mass regularly, and raising his children in the faith. He lived his entire life in Poseyville, rearing his family only a few blocks from where he was born, two blocks from St. Francis, and within walking distance of the new shop he built at 21 N. Endicott Street in 1957.

Illness forced Carl to turn the shop over to Sonny in 1971. Carl died in 1981 at seventy-five and is buried in the St. Francis Cemetery alongside his parents. His wife, Mary, Matthew's great-grandmother, died in 2009 at age 101.

Matthew was born in 1985, so he did not know his great-grandfather Carl, but he did enjoy time with Mary. "She was a wonderful woman, and I was blessed to know her," Matthew said. "She never had a driver's license or a car. I asked her why she did not drive, and she said, 'Honey, why would I need a car? I can walk to the grocery store, and I can walk to church. Everything I need is right here.' That statement has stuck with me. I admire her contentment and that simple way of life."

That simple way of life is still part of the Nix family inheritance. Granddaughter Caroline Nix Eickhoff, who would begin working for the shop in 1987, said, "Grandpa had a passion for using his hands. It was a craft that he turned into a job, and he passed that along to all of us, Sonny, Bill, and myself."[23]

# CHAPTER 5

# Third Generation

## CARL A. "SONNY" NIX JR.: TREAT PEOPLE FAIRLY

Carl A. Nix Jr., known as Sonny, was a child when he started welding in his father's shop, standing on a wooden milk crate to reach the workbench. While he was likely called Sonny because he was the only son, he indeed grew up to be a smiling, jovial man with a "sunny" disposition.

Sonny was born in 1934 and grew up helping at the welding shop, but as an adult, he didn't immediately join the business. He attended Lane Drafting School in Evansville and worked for Mott Wade Construction for a few years, then ran a drill press at the Bucyrus-Erie factory in Evansville. Bucyrus-Erie made steam shovels that built the Panama Canal, bulldozer blades for war use, and draglines that dug up coal in Southern Indiana. He prided himself on having been paid "piece rate" and regularly doing ten or more hours of work in his eight-hour shift.

In April 1957, Sonny, then twenty-three, began working for his father, and he and Carl Sr. constructed a new shop on N. Endicott, laying the block, pouring the concrete floor, and setting the steel roof trusses. On Memorial Day, Sonny welded the decking on the steel beams of the roof. Light streamed in from several sizable sixteen-paned windows that marched around the walls. By fall, they had moved in.[24]

The new home of Nix Welding was described in the newspaper as a "modern" building. It was quite the upgrade from the old wooden shop with a dirt floor packed so hard from years of blacksmithing and welding that when Sonny helped his dad clean up metal shavings and debris, it was like sweeping concrete. The legacy business founded by Charles Nix was incorporated in 1975 as Carl A. Nix Welding Service Inc. The legal entity of the legacy business retains the same name today, though at some point the DBA (which stands for "doing business as") was shortened to Nix Welding. Today the company does business as Nix Industrial, representing the broadening scope of services.

The forty-by-sixty-foot shop with two large wooden garage doors provided more space to work on farm machinery, and it set the course for the next generation of the Nix family to head the business. Carl must have briefly gotten cold feet, though. When the building was complete, he told Sonny, "Maybe I'll just work here awhile to see how things go, and then later on you can join me."

Sonny promptly reminded him, "That's not why you built this shop. You built it for me to come to work with you, so that's what we are going to do."

An old newspaper ad for Carl A. Nix Welding Service depicts a drawing of a man in a welding hood holding a torch and reads, "Carl and Carl, Jr., Owners. 'The Smallest Big Business in Poseyville.'"

While their new building greatly expanded their capabilities, it wasn't all smooth sailing. During their first winter in business together, there was barely enough work to earn a decent living. Farmers were not in the fields and thus were not breaking equipment. One week, Carl and Sonny only had $65 to split between them. Fortunately, they had constructed the new building themselves and were always conservative with spending, so they had no debt. They survived that first slow winter together, and work slowly picked up.

Carl and Sonny earned a reputation within the agriculture community as having the go-to welding shop. Their first "big job" was to repair a wrecked ten-wheeler grain truck, an insurance claim. The truck had flipped over, and the wooden sides of the bed had to be rebuilt. While they only made $640, they felt they had set the world on fire.

Sonny saw several other local welding/machine shops come and go over the years. Most owners lacked the basic inner discipline necessary to sustain self-employment. Others did not treat customers justly or were not dependable. Many failed on both counts. Sonny treated everyone fairly and focused on serving the customer. "When someone would come in with something broken, I never thought about how much money I was going to make," Sonny explained with his signature amiable smile. "I always thought, *This guy is broken down, he's in trouble, and I need to get him back running*."

Working alongside his grandfather and father, Matthew soaked in many lessons about good business and how to foster relationships that still guide him as CEO of Nix Companies. Countless times, Matthew heard Sonny say, "If you do good work and treat people fairly, the rest will take care of itself."

"Although I remember these words," Matthew added, "seeing them modeled day after day was even more powerful than hearing them."

Carl A. "Sonny" Nix Jr., third-generation owner of NIX and grandfather of Matthew Nix. (Photo from *A Nix History* video, 2020.)

A sign near the office read, "Labor rate: $40 per hour. $50 if you watch. $60 if you help." While it was a joke, there was some truth to the sign's poke at customers who tried to micromanage their repairs.

In those days, it was not uncommon for customers to wait in the shop for their repairs to be made. Sometimes a customer would hold one end of a part or do what they could to help speed the process along. One day Sonny was running the hydraulic press, a ram that comes down with seventy tons of force by way of a hydraulic pump and which was used in the shop for straightening bent farm-implement parts. As Sonny ran the press, the motor humming and metal clanging under force, a part shot out and hit a customer in the chest, knocking him to the floor. Had the part hit him in the head, it could have killed him.

Sonny lived in Poseyville his entire life, attending St. Francis Grade School, which was attached to St. Francis Church, and graduating from Poseyville Robb Township High School in 1952. The following year, he married his high school sweetheart, Mary Lee Wilkinson of Stewartsville, a town located a whopping four miles away. Mary Lee was a beautician with a shop attached to the couple's house. In 1954, their first child, Charles William "Bill," Matthew's father, was born. They would go on to have six children. Their fourth child passed away with a birth defect when she was six months old.

Sonny and Mary Lee were a loving couple who raised their children as members of St. Francis Church. Sonny sang in the church choir, accompanied by the organ made possible through the determination of his grandfather Charles. When Bill was about seventeen and his four siblings were quite young, Mary Lee developed a brain tumor that affected her ability to speak. She underwent surgery and treatment, leaving her unable to care for her children. Sonny cared for Mary Lee, raised their children, and ran the welding shop.

Contrary to his emblematic disposition, Sonny was a stern disciplinarian, but given everything on his shoulders, setting rules for his five offspring was essential to keeping his family and life running as smoothly as possible. "When we were growing up, he was kind of hard on us," Bill said. "Every time we messed up, he really got on us. But I guess it all worked out 'cause me and my siblings all turned out pretty good. And I guess I was the same way with all my kids, and they all turned out pretty good."[25]

With rehabilitation, Mary Lee improved and could communicate, but her speech was stilted for the rest of her life. She did not speak often, but when she did, she began nearly every statement with the phrase, "I said . . ."

Mary Lee's brain tumor may have stemmed from a head injury sustained in a car accident when she was fifteen. Three other teenagers in the car, her friends, all died. She was the only passenger to survive.[26]

In the early 2000s, a series of strokes left Mary Lee severely disabled, and Sonny cared for her again. He credited his faith, instilled by his parents, for buoying him to persevere through his difficulties with the positive spirit and smiling face for which he was known. Looking back, he said, "It's a whole lot easier to smile, and if you smile, everybody will wonder what the hell you've been up to. But if you cry, nobody cares."

Matthew has fond memories of his grandmother, for he and his siblings often went to her house after school until their father picked them up when he left the welding shop for the day. Matthew recalled:

> Grandma cooked a huge German dinner each day at noon for Dad, Grandpa, Aunt Caroline, Aunt Carla, and anyone else who might drop by between twelve and one. When I worked at the shop through the summers, I also attended these feasts. Looking back, I'm amazed we could work in the ninety-eight-degree heat of the welding shop, go to Grandma's for a four- or five-course meal, followed by her shaming us into eating cookies, pie, or some other homemade dessert, and then go back to work. I can still see her half-cocked grin and hear her stern voice: "I said . . . have a cookie!"

Each day around her lunch table, Mary Lee served seven to ten people her specialties: pan-fried deer steaks from the venison the Nix men brought home from a day of hunting, pork backbone sauerkraut in a crockpot, and skillet-fried chicken. Bill remembered, "I'll tell ya, we used to deer hunt—and Mom would pan-fry these deer steaks. They were so awesome 'cause hardly no one else could make 'em worth a crap to eat, but she did."[27]

Sonny was a volunteer firefighter called into service during the now famous 1972 Labor Day weekend rock festival held several miles south of Griffin, Indiana, on an obscure Wabash River peninsula known as Bull Island. (The Volunteer Fire Department was located directly across the street from the welding shop.) The festival was said to be Southwest Indiana's mostly peaceful, scaled-down version of 1969's Woodstock in New York State, and it went down in tristate (Indiana, Kentucky, and Illinois) lore simply as "Bull Island."

The newspaper described it as "three days of music, love and peace in a lawless setting where long-haired young people in torn jeans descended like locusts on the dusty farm towns of New Harmony, Griffin, and Poseyville."[28] Beset with legal problems from the start, Bull Island helped close the curtain on the giant outdoor rockfest era.

Sonny, age thirty-eight, was among the volunteers who tried to respond when a mob burned a catering truck at Bull Island. Twenty years later, in 1992, he was quoted in an *Evansville Courier* story about his memory of that day, revealing Sonny's sense of humor.

"We could feel kids jumping on the firetruck as we approached the island," Sonny recalled. "Suddenly, this huge [hippie] wrapped in a blanket stopped us. He looked like a doper. I thought we were in trouble. Then I saw who it was and said, 'Well, John Neidig, what in the world are you doing here?' He said, 'Turn that thing around and get out of here. They'll burn the fire truck.'" Neidig, a fifty-year-old Indiana State Police sergeant who stood six foot eight, was undercover in beads and a floppy hat.[29]

After Mary Lee Nix died in 2003 at the age of seventy, Sonny continued to live a simple life, one that he looked back on with satisfaction and pride. No regrets. "There isn't anyone I can't look in the eye," he said. "I don't have any enemies that I know of."[30]

Sonny went on, "I never felt like I had a job. I always knew what I wanted to do, and I was always glad to be here. I was glad sometimes when the weekend was over so I could come back to work. I always loved my job. I enjoyed every bit of it."[31]

He lived out his adult life one block from the shop that he and his father built in 1957. He walked to work every day, and despite having toiled amid smoke and fumes for more than fifty years, he was blessed with lungs that allowed him to sing in the church choir until his death on December 16, 2022, at age eighty-eight.

Sonny was an avid outdoorsman, golfer, and an exquisite gardener with the undisputed best garden in town, cultivated on a large corner plot next to the shop. The garden was so magnificent that it was featured in the Posey County newspaper.

"I remember working at the shop in the summer as a teenager and being so tired at the end of one of those August days when the heat index was about 105," Matthew said. "While I was feeling sorry for myself, Grandpa would finish his workday, walk up the hill toward his house, take off his sweat-soaked shirt, and start hoeing in the garden."

Today, the land where Sonny gardened lies fallow. Matthew and Adam plan to make it a community garden where NIX employees can plant, tend, hoe, and enjoy the fruits of their labor harvested under the Poseyville sunshine, like Sonny did.

## CHAPTER 6

# Fourth Generation

### BILL NIX AND CAROLINE NIX EICKHOFF: QUALITY AND CUSTOMER CARE

Like his father, Sonny, and grandfather Carl, Bill Nix grew up working in the welding shop. The shop was part of the fabric of their lives, like another family member—joining them at dinner, accompanying them in the Sunday church pew, tagging along on a hunt in the woods near Poseyville.

Bill described learning the ropes: "I had been in there during summer times to learn what to do. You started out with easy stuff, and as you get older and you're around more, and you see how things are done, you figure out how to fix things. The longer you're in there, the more complicated the work gets. After you do it for a while, it all gets pretty easy. But sometimes the jobs that come in get more difficult, too. Everyone knows you can fix most anything, so they think, *Hell, I'll just take it in there to Bill and Sonny, and they'll fix it.* Sometimes you scratch your head for a day or two before you figure out how you're gonna do it."[32]

Like his father, Bill did not join the business immediately after school. He obtained a two-year degree in machine technology at Vincennes University, then worked as a machinist at Babcock & Wilcox (named BWXT today), a nuclear defense contractor in Mount Vernon, Indiana. Then, in 1975, out of the blue, he "told" his father he was coming to work for the family business.

Sonny recalled, "One day he came in and said, 'I'm going to go to work for you.' I said, 'No. You're not going to work *for* me; you are going to work *with* me. When are you going to do that?'"[33]

"Well, I already gave my two weeks' notice," Bill told him. "And it will be up this weekend, so I'll start Monday."

Bill, age twenty-one, started buying in immediately. Carl Sr. still worked in the shop, so the three generations of Nix men welded, repaired, and created solutions to broken machines with their coarse hands marked with grease and dusted with metal shavings. In 1994, Bill would take over the full running of the company, but his father, then sixty, would continue working in the shop.

Three generations, L–R: Carl A. "Sonny" Nix Jr., Carl A. Nix Sr., Charles William "Bill" Nix. (Photo from *Evansville Courier*, 1977.)

The three generations wore the same uniform—steel-blue work shirts with a name patch over their hearts: Carl, Carl, and Bill. Over left pockets filled with pens, soapstone, and a small screwdriver was the company patch, stitched with the words "Carl Nix Welding." Their shirts started out clean in the morning, but by day's end, their bellies, the center of their work togs, would be black from leaning over machinery.

Bill had always thought he would eventually work at the shop. "But you really never know for sure until you get away and see what's going on in the real world," he said. "Then, you come back. I was still living at home when I started working there. It was a block away. You didn't ever drive. You just walked."

Bill's earliest memory of the welding shop is of a day when he was a mere toddler. "I remember when I was about three years old," Bill said in 2023. "I walked from where we lived to where my grandfather had the welding shop uptown. I knew where it was. I walked up there, and somebody had to come pick me up, and I got my butt beat 'cause I wasn't supposed to be up there."

Bill added about Grandfather Carl: "I used to sit and watch him do blacksmith work and run the forge. He hammered out plowshares and bush hog blades. He also sharpened a lot of handsaws with a file and a lot of circular-saw blades and chain saws. He was old enough that he didn't do a lot of physical work, but he was in there fiddling around all the time."

Bill also remembers when he was a boy and Grandpa Carl told him he was going to make some tiny tools—one-eighth scale—using traditional blacksmithing methods. Carl would craft a claw hammer, the most difficult piece, using the forge and anvil. Bill watched with anticipation as Carl began making the promised miniature. But Carl, peering through his wire-framed glasses, began to *weld* a piece of round stock to a piece of square to make the head and claw of the hammer.

"Grandpa, I thought you were going to make those in the forge," Bill said, innocently.

Upon hearing his young grandson's expression of disappointment, Carl threw the pieces in the scrap bucket and started over. This time using the forge and anvil.

"He got him a piece of steel," Bill described decades later. "And he put it in that forge, and he started heatin' it, workin' it, hammerin' it, and stretchin' it, then knocked a hole in it with a square punch."

"I love this story," Matthew said of the family lore, "because it speaks to my great-grandfather Carl's character. He likely forgot he had told my dad how he was going to make the miniature tools, and upon being reminded, he was not going to deviate from his word or disappoint his grandson."

Carl handcrafted a diminutive claw hammer, ball-peen hammer, and hatchet. Decades later, Sonny gave the palm-size tools to Matthew after he became CEO of Nix Companies. Today, Matthew displays the miniatures in his office as a symbol of the Nix dedication to fine workmanship and keeping one's word.

Sonny's daughter, Matthew's aunt Caroline, also remembers watching her grandfather. "Grandpa always told jokes. He was a very caring man. I remember him working the forge back in the corner. When we came down to the shop when we were kids, he always stopped what he was doing and talked to us. He always made us feel very important."[34]

When Bill first started working in the shop, Sonny had to teach him, with tough love, how he and Carl worked. "It was hotter than hell," Sonny said, looking back. "And me and Dad were just busting our butts. I look over, and there's Bill sitting on a chair, running the lathe, reading a hunting magazine. I went over there, and I kicked that chair."

Young Bill looked up at his father. "What do you want?!"

"You get your butt off that chair and run that lathe, or you don't run it at all!" Sonny spat. "We don't sit down while we work around here."

That day, the Nix work ethic was firmly communicated to Charles William "Bill" Nix, who never again sat down to work.

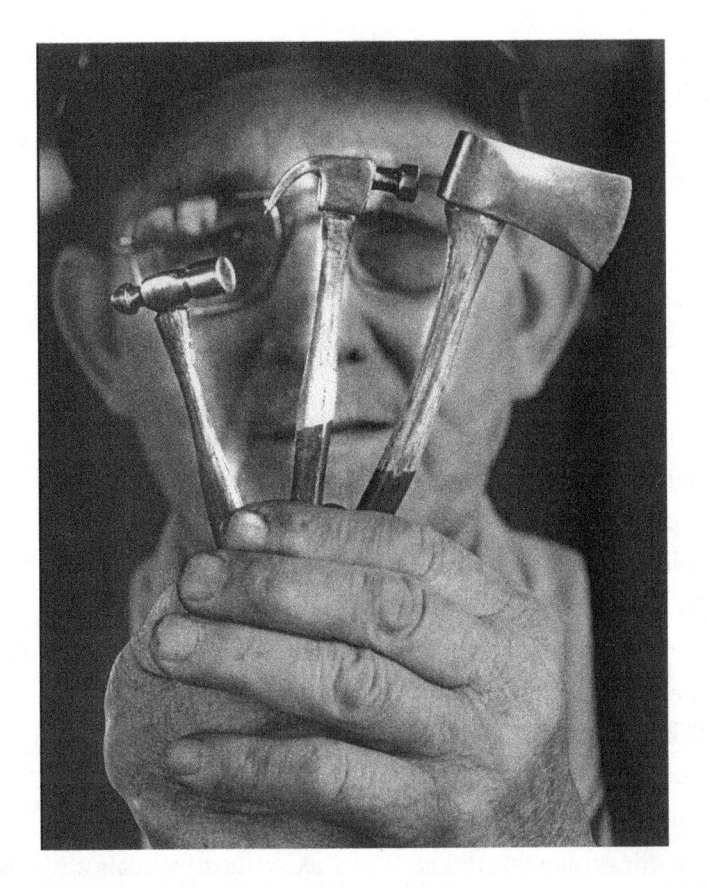

Carl A. Nix Sr. with mini tools he crafted for his grandson Charles William "Bill" Nix. Today, the tools are displayed in the office of his great-grandson Matthew Nix. (Photo from *Evansville Courier*, 1977.)

Bill admits to being a little too fancy-free in those first years. "When we first started working together, it didn't go very good." In describing his younger self's mental state, Bill explained, "You're young and you want to go out and play all night, and you want to come back to work the next day. Some nights you'd go out, and you'd stay out too late, and you wouldn't get enough sleep. And the next morning you had to get up and come to work, and you'd have a frickin' headache 'cause you drank too much the night before."[35]

Bill soon met Donna Marie Blankenberger, a bright young woman with a positive outlook. Her father was a grain farmer. They married on March 24, 1979, at St. Wendel Catholic Church, the church she grew up in about ten miles from Poseyville. Donna went to nursing school and became a licensed practical nurse, working at Welborn Hospital in Evansville as she earned her degree to become a registered nurse the year after she and Bill married. When her four children were young, Donna worked as a newborn intensive care nurse. Bill and Donna still live on their thirty-acre farm three miles north of Poseyville where they raised Matthew and his three siblings.

Donna was introduced to Sonny's quirky sense of humor before she married Bill. She had not been around Sonny much when he asked his future daughter-in-law, "Do you know what 'nix' means?"

"No," Donna said, curious. Was he testing her?

"It means nothing or to stop," Sonny said flatly.

Forty-five years later, Donna recalled, "I think he just wanted my reaction. It's true. Nix means 'nothing.' It was funny."[36]

While Sonny was known for value, Bill was known for quality. And neither attribute is "nothing."

Sonny was of the old school, where cosmetics were not important and anything done that did not add direct value to a repair was a waste; he worked a job until it was "good enough" to save his customer money and get him back in the field to earn a living. He often said, "You aren't building a Rolex. Just get it fixed and get it out the door."

As times changed, the equipment got bigger and more costly, farmers became a bit more affluent, and cosmetics began to matter—a circumstance more suited to Bill's perfectionism. Because of their different philosophies, Sonny and Bill often butted heads on how a job should be done.

Bill explained, "I was more precision and quality because when I went to school, it was for machine work, which is a lot more particular and more accurate than throwing some stuff together and welding it up. With machine work, tool and die, there is not a lot of tolerance. We tried to fix it right so it looked good and it would stay together.

There's crappy work and there's good work. We tried to do good work, and that's what keeps your customers coming back.

"There were people that would have something that was broken, and they would back their truck up to the door, throw it out onto the floor, and say, 'Here, fix it,'" Bill continued. "They didn't care how much it was going to cost. You do your best job, and whatever it costs, it costs. They would be happy with it."

Bill knew his customers, so he did not mince words in his good-natured ribbing when they arrived at the shop for a repair. "People would come in and back up to the door," Bill explained. "And I would say, 'What the hell did you tear up now?' They'd already be in a freakin' bad mood cause they got something tore up; they need it fixed to go back to work. I wouldn't say that to just everybody. No. But people that I knew I could do that to."

Bill remembered a particular job that he believed called for meticulous work that Sonny undertook: "He was restoring an old plow, trying to make it look as well as original. It was a tractor rim, and he couldn't buy the right rim to fit the hub. My dad cut the centers out of the old rims and was putting them in the new ones. I looked at the first one he did and said, 'Are you serious, Dad?' I said, 'That looks like shit. You need to do better than that.' I mean, this guy's trying to restore this old piece of equipment just for looks, and I think you could do better than that. He got all pissed off, and he said, 'Well, here, you just fix the other one.' There was a night-and-day difference between his work and mine. He was always in a hurry to get it done and throw it out the door."

Today, Matthew feels an even greater swing toward attention to detail. He said:

> As the nature of our business has evolved, the customer base has shifted, and the quality standards have increased. Dad is now the "old-timer" griping and saying, "We aren't building a Rolex." While our standards have amplified and much of our work requires a higher level of precision and cosmetic attention

than Dad is used to, his comments aren't always unfounded. The technical term for what he is referencing is "overprocessing," one of the seven types of waste in lean manufacturing.

When Matthew first worked in the shop as a teenager, he often thought he was finished with a job, only to have his father inspect it and tell him he was wrong.

"In the most extreme cases, I would have to start completely over," Matthew recollected. "Since he was paying me by the hour and only charging the customer one time, that meant the business was losing money. But that didn't matter. That never even entered his decision-making. He did what was right, often doing more than what was required, whether it was profitable or not."

"I've always been honest," Bill said. "I don't have a thing to hide. I don't feel bad about any decisions that I have ever made or ever done. You have to be honest and treat everybody the same."

Sonny echoed this sentiment. "There's not anybody that I feel I can't look at face-to-face. I feel comfortable with everybody, and I always did. I always kept a friendly relationship with everybody."

Today, Matthew carries his grandfather and father's work ethic and customer respect into Nix Companies. He said:

Those lessons have stuck with me. Now the dollar amount is much larger. At one point, we had to eat almost $100,000 in warranty between two mistakes we made during about one month's time. That's a big dollar impact for a business our size. In both cases, we could have moved forward by taking a much less impactful approach to our bottom line. In one case, a less ethical approach would have resulted in zero negative impact, but that wasn't even an option in my mind.

We did what was right and dealt with the consequences. If we were owned by investors or publicly traded, we might be influenced to manage for the next-quarter numbers, and that could pressure us into making decisions that go against our values. Since we are family owned and managing the business for the long haul, we can make decisions that negatively impact our financials in a big way in the short term, knowing it's not only the right thing to do but will pay off in the long run.

Four years before Matthew was born, Bill and Sonny built a pole barn next to the cement-block structure his father and grandfather had erected. This building had higher ceilings and large garage doors to allow wider and taller equipment to fit inside. In a continuation of the drive toward growth reflected by Carl's first welding machine and the equipment improvements Sonny made along the way to foster more business, Bill took the leap to add onto the shop and expand services. Had he not done so, Nix Companies would not be the success it is today. The new pole barn was a stepping stone to the future.

In the 1980s, Bill undertook sheet-metal fabrication for small manufacturing plants in the area, building chutes and hoppers, safety rails, and platforms. Bill's sister Caroline Nix Eickhoff joined the company in 1987 to help part-time with the books.

"That lasted about thirty minutes, and I was out in the shop," Caroline said in a 2013 story in *Evansville Business* magazine. Before she started at the shop, Bill did the books at home at night. As the company grew, Caroline began to handle the administrative duties full-time, overseeing accounting, inventory, and purchasing while also working on the shop floor.

Caroline came from a "white-collar job," so her transition to the welding shop was a bit of a culture shock. Wearing a teal-blue sweatshirt bearing the Nix name, blue jeans, and brown leather work boots, she was the only woman at Nix Welding for twenty-five years until Matthew's wife, Lindsey, who holds a degree in management from Indiana University, started in May 2013 to help with sales and marketing.

Sonny and Caroline worked together to create the gate and fencing around St. Francis Xavier Catholic Cemetery next to the church, where generations of Nixes are buried, including Sonny. In his proverbial, humble manner, Sonny gave detail: "Caroline and I painted the gate. The words 'Saint Francis Xavier Cemetery' were cut out with a track torch, a tracer torch. We wrote that out on a big piece of cardboard, and the torch followed the lettering and cut that out of the metal, and it's on that gate out there. If I considered a legacy, that may be as good as I get."[37]

The ironwork is painted a soft seafoam green, and the delicate scrollwork offers a calm and soothing welcome as the gate opens to allow cars into the cemetery. Iron, steel, and metal can be gentle and beautiful if crafted by artisans like father-and-daughter team Sonny Nix and Caroline Nix Eickhoff.

The gate to St. Francis Xavier Catholic Cemetery made by Sonny Nix and his daughter Caroline Nix Eickhoff.

Why was Caroline not brought in as a part owner of the shop as was her brother? Bill said, "Well, we talked about that. She didn't weld. She did some machining work. She could drill holes and lay out stuff, cut parts and pieces. She did a lot of that, but she never did weld anything together, and that's a big thing."

It would seem that Caroline's contributions in accounting, payroll, inventory, and purchasing were not viewed as equally important to her brother and father's hands-on work on the shop floor. How much did the fact that she is a woman play into their thinking regarding her qualifications to become part owner?

Bill described his sister as tough and hardheaded. Perhaps she had to be to hold her own.

Matthew fondly recalled his early days working at the shop and watching Aunt Caroline.

> When she wasn't out on the floor cutting metal or running the lathe, she was in the office doing the bookwork. I rarely saw her sit. Maybe only when she was sending out invoices at the end of the month, which was time consuming and required more concentration. Grandpa wasn't a fan of sitting down at work. She could hold her own with any machinist or bookkeeper. Very few people could do both at her level, regardless if they were male or female. Her administrative work was done from a tiny office with a metal folding chair.

Most of the shop's customers were male, and they liked Caroline because she was a rare breed—someone who could speak their language but had a softer side. She played a maternal role for Matthew and his brother, Adam, shuttling them to athletic practices after school while Bill and Donna were working. This was part of her unwritten job description that she happily performed.

Perhaps Caroline's "soft skills"—relationship building, critical thinking, and various other talents that did not employ the use of machinery—were subconsciously relegated to being part of her "natural role as a woman" and thus not as "important" to the company.

Adam also recalls his aunt with great affection. "We had a really tight bond. She taught me a lot and helped me through tough times. She had a passion for people."

While Caroline was never an owner, her model behavior of how to pinch pennies, be frugal with resources, and serve the customer and the town of Poseyville with fierce loyalty while nurturing family continues to play out in the central values of Nix Companies. She learned these values from her father.

"Sonny is my dad," Caroline states with a gentle expression of admiration in the *Nix History* video filmed before Sonny's passing. "I am very proud that he is my dad. He has taught me so much, working side by side for twenty-five-plus years. He taught me how to be a hard worker, and he would help anybody with anything in any way if he could."[38]

Carl A. "Sonny" Nix with two of his children: Caroline Nix Eickhoff and Charles William "Bill" Nix, circa early 1990s.

Bill's pole barn expansion allowed for two other areas of growth that shaped what would become Nix Companies—semitrailer welding repair and commercial custom fabrication. As a young man, Matthew did many welding repairs on semitrailers. However, the building was not large enough to fit a complete truck and trailer inside.

"If we put the trailer in and shut the door, it was so tight in the shop you could barely walk or pull the welder around it," Matthew said. If a trailer needed a quick fix, it didn't make sense to go through the trouble to unhook the tractor from the trailer. When it was cold outside, a trailer was backed in and the garage door pulled down between the truck cab and the trailer to keep as much heat as possible in the shop. Lighting was minimal. When working under the trailer, a portable light or a headlamp was needed. Additionally, although Bill and Sonny thought they had built the barn big enough to work on trailers of any size, box trailers were 13.5 feet tall and did not fit.

Despite the nuances of the pole barn, it was a sufficient space to allow for an astonishing amount of work to take place. Many times, as soon as one job pulled out, the next one pulled in. Often they were waiting three deep.

The welding shop began offering commercial custom fabrication thanks to the business provided by two small plastics plants in town: All Tech Plastics and their offshoot, Hoehn Plastics, which is still family owned and has been a loyal customer of NIX for two generations. Locally-owned All Tech eventually consolidated to Evansville. In the spirit of all things connected, its former building on Frontage Road was sold at auction in 2016 to Nix Companies. Today, a renovated version of the facility is the company's central operations, an impressive building surrounded by farm fields. The architectural drawing is complete for a new headquarters office with a soaring glass atrium that can be seen from Interstate 64.

The addition of a track torch, handheld plasma cutter, sheet-metal brake, and roller allowed the shop to do a wide range of sheet-metal and metal fabrication projects for the two plastics plants and other local businesses. Fabrication projects such as hoppers, chutes, transitions, stairs, handrails, and platforms are still part of Nix Companies' primary business. The equipment and technology deployed to create the custom metal have evolved, but the types of projects are largely the same.

Custom fabrication was attractive to Matthew. While repair work provides instant gratification and a sense of fulfillment, Matthew was most drawn to the idea of creating something from nothing. Matthew never stopped dreaming, fabricating possibilities for the company that his great-great-grandfather Charles Nix started as a simple blacksmith shop.

"Without the shop expansion, I don't know if I would have joined the business," Matthew admitted. "And my brother, Adam, certainly would not have come on board in 2011. Each generation left its mark. The reputation for quality work, the advancement of capabilities, and the pole barn addition to the shop were certainly Dad's."

However, Bill gives much of the credit for the pole barn to Sonny. In 2023, Bill explained, "I'm really not much of an expander. I'm nothing like Matthew. My dad was the one that really thought we needed to build on because we were working on large equipment outside in the wintertime and it was freaking cold. We built a building as big as we could for the property that we were on, and some of the equipment still would not come in the doors. But it sure did make a big difference. It [working conditions] was a lot better than it was."

The warm air from two heaters set at fifty-five degrees in the original cement-block building flowed into the pole barn. No additional heater was added to the new structure. "We did add some air cleaners in there, 'cause when you weld, you get so much smoke," Bill said. "Then instead of opening the doors and turning on a fan to suck all the smoke out, the air cleaners purified the air. So you didn't lose all your heat, which helped a lot."

Caroline summed up the Nix way: "If a customer needed something fixed, it wasn't about the money. It was about taking care of the customer, having great customer service, and that's what we provided."

"Sometimes you *are* building a Rolex," Bill said. "Anything that came through the shop that was very particular, that had to be done well, I mean *right*, then I did a lot of that. It also depended on the customer. Some guys were a lot more particular than others. You got to know when you're building a Rolex and when you're not."

## CHAPTER 7

# Fifth Generation

### MATTHEW NIX AND ADAM NIX: LOVE-HATE RELATIONSHIP WITH THE SHOP

Like their father, grandfather, and great-grandfather, Matthew and Adam Nix grew up sweeping the floor of the shop and learning to weld, make minor repairs, and handle fabrication jobs. Their sisters, Jennifer and Lauren, were born in 1990 and 1991, respectively, and never worked in the shop. Today, Jennifer is a kindergarten teacher at Saint Wendel Catholic Church, where her mother attended, was baptized, and married her father. Lauren is a nurse by trade like Donna but now works on her husband's family farm. The Nix women hold their own legacy of career connections.

While Adam hated the work, Matthew, three years his brother's senior, enjoyed much of his time working in the shop on Saturdays and during the summers when in high school, even some of the miserably hot days filled with backbreaking labor.

In the 1990s, the brothers were paid $5 for a half day's work cleaning up the shop. In 2000, when Matthew was fifteen, Indiana's minimum wage was $5.15 per hour. Eventually, Matthew received a raise to $20 a day. The first time he worked full-time for a five-day week, he made $100 and thought he was rich.

As a teenager, Matthew learned how to use some of the tools in the shop. Most of the time, he cut material on the band saw and the ironworker and drilled holes using the drill press. An ironworker is a machine that shears, notches, and punches holes in steel plate. He would eventually run the lathe and milling machine. He enjoyed working with his hands and creating.

The Nix family in 2006. Standing, Donna and Matthew. Second row, Bill and Adam. Bottom row, Jennifer and Lauren.

"We made a lot of U-bolts in those days for farm repairs," Matthew said. "This was the worst job on a hot summer afternoon because the threading machine sat in front of one of the big glass windows in the block building, facing the west. There was no air movement in that old building, and it must have been 120 degrees with that sun baking down. Getting the machine set just right to cut the threads deep enough that the nut would thread on easily but not so deep that it was loose and lost its integrity was a finicky process."

Sweat dripping down his face, Matthew would thread with well-used dies. "I distinctly remember being ringing wet and looking through sweat-covered safety glasses, trying to get the dies changed out and adjusted. If there were ever times that I wanted to get away from the welding shop, it was then."

Sonny well remembered the Indiana heat. "We worked hard. We used to wring water out of our underwear when we got through in the summertime."[39]

Matthew learned to weld by building up tractor "plow points" that had been worn down from their time in the dirt. "It's almost impossible to mess up this job," Matthew said. "The buildup process is time consuming, so it was the perfect way to learn. I was earning money for the business by burning rods or wire and getting tons of practice stacking beads on top of one another. After the buildup was complete, Dad or Grandpa would apply the hard surfacing with special welding rods. Eventually, I would learn to do that as well."

A weld bead is created by applying filler material to a joint between two pieces of metal. The process creates a lot of smoke, and with the garage doors shut in the wintertime, a fog hovered in the air, thick and ominous.

As the years went by, Matthew earned the right to do minor repair and fabrication jobs using the wire welder. By the time he was in high school, he "MIG-welded patches in aluminum dump trailers and TIG-welded aluminum fuel tanks." MIG stands for "metal inert gas" and TIG for "tungsten inert gas."

He worked in the shop every summer throughout his teenage years except for one. "When I was a junior, I got a job with a friend, building wooden lake docks," Matthew said. "Even though that job paid better per hour than what I got paid at the shop, they didn't always have enough work to keep me busy. Once Dad found out I wasn't working forty hours, he made me work at the welding shop during my downtimes. That was the first and last job I've ever had working for anyone other than my dad or myself."

Adam's childhood relationship with the welding shop presents a striking juxtaposition with his current position as one of three vice presidents of operations for NIX.

"The welding shop was always used as a punishment tactic for my mother to impose on me when I was being difficult at home," Adam reminisced in his soft-spoken manner. "Mom couldn't physically punish me enough at home to get me to straighten up, so she'd resort to this, and it worked every time. 'I'm going to send you with your dad to the welding shop!' I still remember her saying those words. Due to this, my early memories of the shop aren't as fond as Matthew's—although, looking back, I appreciate the time I was able to spend with Dad, Grandpa, and Aunt Caroline. They helped shape me into who I am today."

Adam has fond recollections of his father letting him use the MIG welder to "tack and splatter" metal scrap pieces together when he was twelve. "I thought I had set the world on fire with my newly acquired talent. So Dad and Grandpa began to call me 'Sparky.'"

When he wasn't playing baseball, Adam spent summers working in the shop—until 2004, when he was sixteen. That summer was a learning experience in the satisfaction of putting in a hard day's work. He was a lifeguard at Harmonie State Park in New Harmony, an idyllic, historic town along the Wabash River about fifteen miles from Poseyville.

"I was able to sleep in, as my shift didn't start until 11 a.m.," Adam said. "I worked until 6 p.m. The job was pretty simple. I'd spend forty-five minutes in the lifeguard chair and then get a forty-five-minute break. Although I thought this was ridiculous, it beat the heck out of starting at 7:30 a.m. and working to 5 p.m. and physically laboring every ounce of every day at the shop. However, over the course of that summer, I felt lazy, and I wasn't ultimately fulfilled. I felt guilty for the relative ease of that job."

The lifeguard gig paid well, more than Adam's friends were paid to work in the nearby melon fields or do construction work. He had the time and energy to spend with his buddies on June nights, staying out late because he didn't have to be at work until midmorning the next day. Adam was paid to "sit down on the job" in the lifeguard chair. This ran counter to the mantra Grandpa Sonny touted in his welding shop. No one sat down on Sonny's watch.

The following summer, Adam had the option to again be a lifeguard; however, he decided to return to the shop. His conscience could not ignore the work ethic instilled by his grandfather, father, and aunt Caroline.

"I found it meaningless," Adam said of his lifeguard job. "I decided to go back and work in the shop, even though I still didn't like it at that time. Now I'm glad I did."

Adam admits that he had a "rebel mentality" and was a bit of a "hothead." When he was a sophomore, he won the starting position on the football team as a cornerback. "Not due to my ability," Adam said, "but to the fact the coaches didn't have anyone else to choose from." Smaller than the other players, he struggled. Bill gave him some strategic advice: "You're not going to be bigger than them, but you can be meaner and tougher than them."

Those words stuck with Adam. "I believe in hard work, grit, and a positive, winning, whatever-it-takes mentality," he said. He carried that approach into Nix Companies. "I now cope better when things don't go our way if we learn from our mistakes and ensure we don't continue to repeat those mistakes."

Adam attended Vincennes University, majoring in business administration. He also took a night class in welding at his father's request. "To be frank, I didn't learn much from the class due to the experiences I'd gained during the summers with Dad, Grandpa, and Matthew, but I did get a lot of repetition laying various beads for the different welding processes."

Every student had to take a public speaking class. "Public speaking ranks near the top of everyone's fears, and I was no different," Adam said. "I learned to appreciate it, though, and to this day, one of my favorite things to do is get in front of the folks in the business and speak to them during our weekly meetings."

After one year at Vincennes, Adam transferred to the University of Southern Indiana in 2007 and majored in marketing. "Looking back on my higher education, I can say I acquired two important life lessons. One, it taught me how to learn. And second, whenever you're working through a tough problem, always work it easiest to hardest. I learned this from a philosophy class that at first I thought was a ridiculous requirement for graduation but now value. As you work through the easier issues of a problem, when you get to the more difficult aspects, they don't seem as hard."

Matthew's path was more straightforward. When he was a senior at North Posey High School, he had enough credit hours to attend school for half a day and work the other half at the shop. After graduation, he attended Vincennes University's one-year program in welding, an immersive course combining classroom work with four-hour labs. "More computer classes or a business degree would have served me much better than a welding certificate," Matthew said in hindsight. "But at the time, I had no way of knowing what our business would look like today, and my aspirations for growing it had not yet entered my mind."

After earning his certificate and becoming certified as an American Welder Society welder in 5G pipe, Matthew started working full-time at Nix Welding in May 2004, when he was nineteen.

However, Sonny and Bill didn't think much of Matthew's welding certification. "My dad and grandpa mostly minimized the certification and said, 'You know what you can do with that piece of paper? You can wipe your ass with it.'

"Although looking back I really value my time working with my dad and grandpa, and we have a great relationship today, it wasn't always the most positive or uplifting environment. They had a way of keeping you humble. Some would even call it 'beating you down.' I would tell the young guys who worked around my dad, 'Don't expect any pats on the back from him. If he's not chewing your butt, you are doing a good job. And if he ever quits chewing you out altogether, that means he doesn't like you, and he's given up on you.'"

Bill explained his negative comments about Matthew's certification in an interview—perhaps an explanation that Matthew has not heard. Bill said: "When you do a welding certification, on that given day, you do one test, and everything is perfect. The cut and fit are perfect. It's all sitting there on the welding bench. You can get around it and do everything you need to do, and yes, you passed that test one time on that day, when everything's perfect.

"But when you're working doing what we were doing, nothing is ever perfect. Nothing. You're crawling underneath something, you're laying on your back and you're stretching, and you can't see what the hell you're trying to weld. The damn sparks are running down your neck and burning your back and your arms. It's totally different when you're trying to do that on the welding bench, trying to pass that certification.

"People who pass the certification test, yes, they can weld, but conditions control how well you do that. That's what I was talking about with Matthew's certification."[40]

Nineteen-year-old Matthew would take his certification and run with it, expanding Nix Welding beyond Sonny and Bill's wildest aspirations—beyond shimmying under a low-hung tractor and enduring sparks flying down one's shirt collar. Yet always with a respectful nod to his family's noble welding legacy.

# PART 2

# Fifth Generation Finds the Way

**CHAPTER 8**

# Paid to Figure It Out

Much has changed since Matthew started working in his father's welding shop full-time in 2004, yet much remains the same. Back then there were four employees: Sonny, Bill, Caroline, and Matthew. Work was based on a Sonny Nix philosophy that remains foundational to Nix Companies: "We're paid to figure it out."

The office in the cement-block building where Matthew's aunt Caroline did the administrative work was the size of a closet. With barely enough room for two people to stand, it held an old wooden desk, a metal cabinet, and a metal chair used chiefly as a step stool. The shop's first computer arrived in 2003, compliments of Matthew.

"I won a computer as a door prize at my high school senior after-prom," Matthew recalled. "My parents had already purchased a laptop for me for college, so I gave that Dell desktop to Aunt Caroline and said, 'Here, put this in the office.' She had wanted an office computer for years, so she happily obliged. Unfortunately, that would be nearly the only change of the many I would foster in the following decade that she and I would agree upon."

The welding bench served as a conference table. "There were many meetings around that workbench," Matthew said. "Most were to draw makeshift blueprints with soapstone or review a bill of material or job instructions. However, workbench meetings were not always project related. If we needed to have a family discussion, it happened around that workbench."

Discussions often centered on family, friends, small-town news, and area sports. Adam had a legendary athletic career for the North Posey Vikings. He was the first and only athlete to go to the state championship in the same year for three sports: baseball, wrestling, and football. In fact, he went to state four consecutive seasons: for junior baseball, senior football, senior wrestling, and senior baseball. That was a point of pride for much of the family workbench-turned-conference-table banter in 2006. In 2019, Adam was inducted into the North Posey High School Athletic Hall of Fame.

Back then, the bill of materials and work instructions for Nix Welding were written on small pieces of paper that Caroline recycled from used printer paper and cut into small squares. Nothing was wasted. "Grandpa was so frugal and so against waste," Matthew explained. "If you threw a paper towel in the trash can [an old five-gallon bucket] after washing your hands and Grandpa didn't think it was wet enough, he would reuse it to dry his own hands instead of getting a new one."

Matthew Nix, age nineteen, in the welding shop, 2004.

Many arguments erupted regarding the size of scrap metal that Matthew and Adam threw away. "There was no set standard size that was acceptable to pitch," Matthew said.

> The minimum size to keep depended on who was overseeing you—Dad, Grandpa, or Caroline—and what kind of a mood they were in. It was not uncommon to be reprimanded for throwing away a piece of one-quarter-by-two-inch flat bar that was twelve inches long. The cost of this size of scrap would be about fifty cents. I understand not throwing away fifty cents, but the problem is that piece of steel took up one hundred times more space than two quarters did, and we spent one dollar in labor looking for the steel that was saved.

Matthew understands and appreciates the principle and the history that shaped his family's frugality. He believes in living by one's principles. "But I also believe that when scientific or mathematical facts prove your convictions wrong, it's time to adjust course and reevaluate standards of practice." This ride-the-tide-of-change philosophy helped Matthew magnify the company into a burgeoning national brand.

"Today, we scrap pieces worth hundreds of dollars," Matthew admitted. "And it makes me cringe. My early days in the shop still influence me. But if we kept it all, we would need ten acres of laydown yard and a full-time person just to keep track of it all. When you look at the big picture, it doesn't make mathematical sense to keep it. Additionally, with technology and improved processes, as a percentage of total steel purchased, we 'scrap' [recycle] far less steel today than we did when I started full-time in the shop, so it doesn't make emotional sense to get hung up over keeping every piece."

During Matthew's early days at the shop, a customer once backed his truck up to the garage door with a piece of equipment so mangled it was unrecognizable. Matthew can't remember what the equipment was, but he vividly remembers the lesson he learned from Sonny when he helped his grandfather unload the contorted piece. "I had worked around Grandpa for years, and I could sense he wasn't exactly sure how he would repair the part."

After the customer left, Matthew asked, "Grandpa, do you know how to fix that?"

"Nope," Sonny said.

"Then why did you take it in? Why didn't you just tell that customer you couldn't fix it?"

Sonny looked at his fresh-faced grandson. "Matthew, we don't get paid to say no. We get paid to figure it out."

This was a lesson in "You can't just take the easy jobs."

"Sometimes it takes the pain-in-the-ass jobs to help pay the bills," Matthew said. "And those are often the jobs that earn us the most merit with our customers and lead to future, better projects."

Today, one of the savvy salespeople at Nix, Corey Wilzbacher, asks new customers, "What is something that's a real pain in your ass that no one else has helped you solve?" This is a great strategy to build customer trust and loyalty. While it can lead to an "undesirable" project, as long it is not outside the company's capabilities, Matthew believes the "unsolvable" job should be tackled, even if it stretches the NIX team out of its comfort zone.

Another lesson in frugality and customer service occurred when Bill and Sonny were having "one of their usual bickering sessions" regarding a repair project. "Dad, being the more quality focused of the two, and Grandpa, clinging to his frugality, were not going to agree on the balance between value engineering the solution and delivering a top-notch finished product superior to when it was new," Matthew surmised.

Sonny was the one who had communicated with the customer and so believed he could better speak to what the customer wanted. He looked at his son and said, "Bill, I don't care how you want to do it. We are going to do it the way the customer wants. If the customer tells us to piss on it, we will whip it out and piss on it."

Today, Matthew uses a version of that same comment with members of his team, censoring the phrase according to the audience or context in which he is "quoting" his grandfather.

Their customers' esteem for the Sonny Nix philosophy on impeccable, honest, can-do service was evident during the holidays. "We were flooded with gifts from our customers," Matthew said. "Everything from fruit baskets to hams, salami, turkeys, and gift cards. Today, I use this as an example for our sales team. They want to give our customers gifts. We do, and I acknowledge it's a nice gesture of appreciation, but I tell them, if we take great care of our customers, they will be the ones buying us gifts."

One day, the phone rang in the shop, and Bill answered. It was Mr. Williams, a Posey County farmer. Bill hung up and said, "Matthew, I want you to go on this repair run."

Mr. Williams had a piece of tillage equipment with a broken frame. Bill reviewed the scope of the repair with his son and sent him on his way. It would be Matthew's first solo service call.

"I arrived at the job site, a dirt parking lot next to an old equipment shed on a county road," Matthew said. "I assessed the damage and compared it to my mental notes from Dad's explanation of the problem and how he would tackle it. You can rarely predict the way a field job will unfold, so you have to make real-time adjustments and problem-solve on the fly, which is part of what makes the field interesting and fulfilling. This particular day was no exception."

Mr. Williams, a seasoned farmer, watched intently as young Matthew evaluated the fix and adjusted his strategy. Matthew could sense Mr. Williams's confidence in his ability to complete the repair dwindling by the moment. "He began to question every move I made and my approach to the repair," Matthew said. While Matthew never saw himself as the best welder, his problem-solving skills were his strong suit, and he knew it, even in his early twenties. This inner knowing gave Matthew the chutzpah to stand up to Mr. Williams, a man in his seventies.

"I knew intuitively that he would not hold what I had to say in much regard, so I attempted to take control of the situation by confidently saying in the most polite and respectful way, 'Mr. Williams, my dad has been fixing farm equipment his whole life. If he didn't think I could handle the job, he wouldn't have sent me.' And just like that, his tone changed. He quit questioning my judgment, and I got on with the repair."

A few weeks later, Bill told Matthew that he'd gotten another call from Mr. Williams. This time it was to tell him that he was impressed with Matthew's repair and with the way he handled the situation. Matthew's first solo service call was a success. He had pleased his father, and that alone was worth everything to Matthew, but he also learned that respectfully articulating confidence in his own abilities could lead to a powerful outcome. It could change the tenor of a customer interface from negative to positive.

While Matthew has fond memories of working alongside his father, and the lessons Bill taught him are numerous, the two were often at odds. "Partly because he was a hard-nosed boss and tends to be gruff with people," Matthew said. "And also because we have totally opposite personalities and views on business, which put us on a collision course."

The longer Matthew worked and observed the business, the more his aspirations for growth increased, and the more pressure it put on their professional relationship. Bill's philosophy was "Keep it simple," while Matthew had a natural ambition for more. He wanted to act, make things happen. However, their disagreements rarely extended beyond the shop. "We were always good about leaving work at work and continuing to hunt or fish and have Thanksgiving together as father and son," Matthew said.

The family had not done much marketing for the shop. Matthew suggested they have the company name and phone number painted on the side of the service truck. Bill and Caroline did not see the need. "Everyone already knows our phone number," they said. "Why would we need to put it on our truck?"

"Their argument was ridiculous, and their reaction still astonishes me," Matthew said decades later. "But it speaks to how narrow their focus was."

Bill had a phrase he liked to use when he was about to do something you told him he couldn't: "Well, you just hide and watch." Matthew soaked in his father's words and developed his own go-to phrase: "Ask for forgiveness, not permission."

Many times over the next few years, Matthew would utilize that philosophy, beginning with the truck signage. "I ordered the signs myself and affixed them to the sides of our truck," he said.

Shortly after, Matthew was in the truck on a service call when another truck pulled up alongside him. The driver was with a utility company bringing the first broadband internet to the area. He asked Matthew what kind of work Nix Welding did.

The driver told Matthew that the utility company would be doing horizontal drilling, conventional excavation, and other infrastructure construction and would need field welding, as well as shop repairs and fabrication. "And indeed they did," Matthew recalled. "Over the next couple of years, they became one of our best customers. Those seventy-dollar decals I placed on our truck yielded thousands of dollars in new business. I stood up to the old-school way of thinking, and it paid off."

Matthew's small idea garnered big results, and a seed was planted. He began to intentionally focus more on sales and marketing to cultivate business growth. However, the family pushback continued. Matthew made business cards, developed a website, and added additional phone lines in the face of Bill and Caroline's resistance. Each addition was a modest investment that quickly proved to be profitable.

Years later, Bill explained, "When my dad and my sister were working, we were busy all the time. We never had to worry about work, and we never had to advertise because people always found us."[41]

Now retired, Bill looks back with perspective and praises his son. "When Matthew came in, with another hand, in the wintertime it would get a little slow, and he thought we needed to get more work. He was young and energetic, and he's a worker. He has a hell of a work ethic."[42]

Before the advent of cell phones, when the shop had only one phone line, customers had to call four or five times before they could get through.

When Matthew was out on a service call and phoned the shop, he had to call several times before the line was not "busy." This was *not* good customer service, and in Matthew's view, he was "paid to figure it out."

"It's hard to imagine the amount of business we were losing or that anyone would dispute the thirty-dollar-a-month investment for a second line," Matthew said. "But in order to make the change to more phone lines, one would first need to *want* more business. That is precisely where the divergence was rooted. I wanted to grow; Dad and Aunt Caroline wanted to stay the same."

Matthew fought and ultimately won the family clash between remaining status quo and transforming for advancement. Being the only voice for change was lonely, but it built character, forcing Matthew to be scrappy and make smart decisions. He honed his skills to argue for what he believed in and learned early on that "you never stop selling."

"Whether it's selling the old guard on new ideas," Matthew explained, "selling to new customers, selling to top talent to join our team, or selling stakeholders on fresh concepts, we are always selling."

Soon, Matthew, at age twenty-three, would convince his father to sell him half the company—a shabby building and a few timeworn pieces of equipment, with a job thrown in. To an outsider driving through Poseyville, it might have looked like a bad business decision. What could this young whippersnapper do with an outdated, rusty old shop in the middle of nowhere?

Well, you just hide and watch.

# CHAPTER 9

# Maternal Combustion

Matthew was working in the shop on a blustery late November day in 2006 when he looked up and saw his mother walk in. Where Matthew describes his dad as having a good heart yet "rough around the edges and sometimes rude," he describes Donna as "warm and friendly." This day, as she stood in front of her son on the steely cold floor of the shop, a long shadow fell over her customarily congenial countenance.

"I could tell by the look on her face something was wrong," Matthew recalled. "The first thing I thought of was that one of my siblings had been in a car wreck, because she looked like she had seen a ghost."[43]

Indeed, the ghost of their collective past had phoned.

"St. Elizabeth called. It's your mom," Donna said.

Matthew stared at his mother. *What? You are my mom*, he thought.

"She's sick, and her family wants to connect with you."

Matthew and Adam knew they had been adopted from what is today St. Elizabeth/Coleman Pregnancy and Adoption Services, part of Catholic Charities Indianapolis. Donna had communicated Matthew's adoption story as a natural fact of his life from the time he was old enough to comprehend. After they adopted their sons, Donna and Bill had two biological daughters.

"I grew up with the stereotype in my mind that my biological mom was probably poor or a drug addict," Matthew explained years later. "With that image, I felt thankful that she chose life and gave me up for adoption."

The next forty-eight hours after Donna visited the shop and dropped this bombshell were crazy. Matthew's life shifted in that pivotal moment of declaration from his adoptive mother, chosen as the one to raise him twenty-one years earlier by his biological mother. Maureen Patrice Williams Maddox and her family—his blood family—were about to become real to Matthew.

Donna and Bill lived in the same house they were in when they adopted Matthew, so they still had the landline phone number that St. Elizabeth had on file. Donna had returned from work at the neonatal intensive care unit at the Women's Hospital in Evansville, where she cared for sick and struggling infants, and found a message on the answering machine. A representative from St. Elizabeth explained that Matthew's biological mother was ill, and the family wanted to contact him. If Matthew was agreeable to connecting, he was to call St. Elizabeth. He and Donna began scrambling to track down more information.

Meanwhile, unbeknownst to Matthew, his biological grandmother Jean Marie Williams, a registered nurse who was a stalwart Catholic and, on happier days, fun loving and witty, was desperately trying to get the closed adoption records released. Maureen was dying of colon cancer at forty-one and already floating in and out of consciousness. Days earlier, she had told her mother that before she died, she wanted to see the son she gave birth to on April 13, 1985, and never saw again. Jean made it her mission, come hell or high water, to honor her dying daughter's wish to see the son she gallantly, with deep faith, gave away.

Never underestimate the power of a mother trying to do right by her children. Maureen learned that from Jean. And Jean learned that from Maureen.

Jean was about to lose a daughter. And gain a grandson.

"My biological grandmother somehow tracked down the judge on a Saturday because she needed the legal paperwork," Matthew said. "And the judge said, 'Do whatever you need to do. We'll worry about the paperwork next week.' Then he released the information to her. Which was basically against the law, but given the circumstance, he did what was right. I think it's neat that my biological grandmother had the drive to do that."

Matthew obtained Maureen's address and was able to phone her husband, Tim Maddox. After hearing the news, Matthew's girlfriend immediately drove to his house in Poseyville. Lindsey Tenbarge, who would later become his wife, was a student at Indiana University, and the two had been dating since they were both fifteen. Matthew asked Lindsey and Donna to accompany him on the three-hour drive to Franklin, Tennessee, to see his biological mother. He brought a bouquet of flowers.

"We pulled up to their address, a really nice gated subdivision outside of Nashville," Matthew said. "It was interesting. I had this vision my whole life of her being underprivileged or wayward. But in reality, she had a career in human resources, and she married a guy who was in finance. They had done well and built a nice life together."

Matthew walked into the home and saw Maureen lying unconscious in a hospital bed. "I held her hand and talked to her," Matthew remembered.

Donna watched her son meet his biological mother. "She never talked, but she was definitely listening, and her husband was so nice to allow Matthew in the home," Donna said. "It was sad, but it was good for Matthew."[44]

Tim talked to Matthew, Donna, and Lindsey about his life with Maureen. "They had a lot of the same interests that our family has," Donna said. "We both spent summers boating on the lake with our kids."

A couple of hours after meeting Maureen's family, *his* family, and talking to his biological mother for the first and last time, Matthew climbed into his car and drove home. Before he arrived back in Poseyville, the town he had called home for an entire life lived without her, Maureen Patrice Williams Maddox died. "I like to think that she knew I was there by her bedside, and she was waiting for my visit before she passed," Matthew said.

A few days later, Matthew returned to Tennessee for Maureen's funeral. His aunt Caroline and Lindsey came along. Donna did not attend the funeral. "She didn't feel like it was appropriate," Matthew said. "It was a lot for my mom to process, too."

Matthew's mother Donna Marie Blankenberger Nix in 2007 when she was the school nurse at North Posey High School, where all of her children attended.

Matthew met more of his biological family at the funeral and learned the backstory of his birth. While attending Indiana University, Maureen had a "boyfriend fling" and became pregnant. She dropped out of IU and returned to Indianapolis, where her parents lived and Matthew was born. Matthew still does not know anything about his biological father. Maureen's family has offered to help connect him, but he isn't ready to go down that path again. At least, not at this point.

Maureen Patrice Williams Maddox, biological mother of Matthew Nix, as she looked about the time that she gave birth to him at age twenty and placed him for adoption.

Maureen married Tim Maddox in December 1986. They moved to Lafayette, Indiana, for Tim's job at Arnett Clinic, and Maureen finished her degree at Purdue University. Ironically, later they were unable to start a family and adopted two children.

Matthew is Maureen's only biological child. He remains connected to her family and feels at home with them. "When I first came to family gatherings, it was weird," Matthew recalled with a laugh. "They would say, 'Oh, you look like Maureen.' They are just wonderful people.

They're a devout Catholic family. We are, too. That was why they went to St. Elizabeth. And the reason why I think Maureen chose my adoptive parents.

"I didn't really ever have a strong desire to track my family down, but it was in the back of my mind as I was growing up," Matthew said. "You had to be twenty-one before adoption papers could be released in Indiana. I had not filled out the paperwork, but the thought had crossed my mind. I just never got around to it."

Matthew mourned his biological mother's passing, but having not met her until the day of her death, what he grieved most was the loss of the *chance* to know her, ask her questions, learn from her, hug her, and feel her hug in return. The window of opportunity to establish a relationship opened, then shut on the same day. That loss caused a seismic shift in Matthew's world.

"It was the only time in my life where I was evaluating everything and turning everything upside down," Matthew said. The first aspect to turn topsy-turvy was his relationship with his longtime girlfriend. He broke up with her the summer after Maureen died. Even though he had already saved the money to buy her an engagement ring.

A strong, clear-thinking, intuitive, raven-haired young woman, Lindsey Nicole Tenbarge of Haubstadt, another small community with German roots not far from Poseyville, had been a high school cheerleader, raised to be "super independent" as one of four daughters. Today, she is director of public relations and training for Nix Companies.

"Matthew questioned a lot of things about his life at that point," Lindsey remembered. "He asked himself, *Is this what I'm supposed to be doing? Is this who I'm supposed to be with?*"[45]

Lindsey had broken up with Matthew when she was a freshman at IU and later got back with him, so her thought was *Well, this is his time to take a breather from the relationship.* "He was just off; there was something off about him. He had to process it," she said.[46]

Lindsey headed into her senior year at IU in fall 2007, just after the breakup. "Actually it was a blessing," Lindsey explained. "Because then I did all the senior things with my girlfriends in my sorority and didn't leave campus every weekend to come home and see Matthew. I had a lot of fun."[47]

Matthew began dating another woman, who was, Lindsey said, "very different than I am." It was a hard time for Matthew's parents, too, as they watched their son navigate the new terrain of his emotions. Donna reflected, "Matthew went through the only bad time I've ever seen him go through where out of the blue he broke up with Lindsey. He had met this girl, and his sisters and I did not like her. We did not like how she presented herself. Matthew knew we did not approve of her and that we loved Lindsey.

"During this time, I was just devastated. And Bill, too. We would talk and say that it had something to do with Matthew finding his mom. And now she's gone. Or maybe it was because he was thinking about getting married and that can go to the head. It's hard to say. Some people get cold feet before they think about marriage, and maybe that was it, because they were married within a year or two after that."[48]

While his love life took a turn, Matthew was knee-deep in a side hustle he started—the Outdoor Connections, Inc., where he sold hunting equipment online. The business morphed into a store and organized game hunts for the public. "It was a side business I ran for a few years as I scratched my entrepreneurial itch before Dad would let me run with the welding business," Matthew said.

In hindsight, the Maureen experience appears to have kicked him into high gear to search for more in his career. Before the fateful call from St. Elizabeth, he thought he would work in his family's welding shop for his entire life, just like his father and grandfather and all the Nix men before him. Maureen's death was like gasoline on the fire of Matthew's budding innovative spirit. Like the gasoline that ignited the kindling in his great-grandfather's forge.

Although he did not recognize it then, and still appears to *not* see Maureen's death as the catalyst to his change, Matthew's pumped-up drive very much resembles the drive he admired in the biological grandmother who fought doggedly to have a judge release adoption records so she could find him before her beloved daughter died.

One of the significant books Matthew credits for inspiring him to build Nix Companies is *Good to Great* by Jim Collins. A number one bestseller when it was published in 2001, the book is about "why some companies make the leap . . . and others don't." Collins writes that every good-to-great company covered in his book had what he terms "level 5 leadership" during pivotal transition years. He pens:

> Level 5 leaders embody a paradoxical mix of personal humility and professional will. They are ambitious, to be sure, but ambitious first and foremost for the company, not themselves. Level 5 leaders display a compelling modesty, are self-effacing and understated. [They] are frantically driven, infected with an incurable need to produce sustained *results*. They are resolved to do whatever it takes to make the company great, no matter how big or hard the decisions. [They] display a workmanlike diligence—more plow horse than show horse.[49]

Words in Collins's book speak to Matthew's "Maureen moment" that caused his existential crisis. Collins writes that leaders with the potential to evolve to level 5 have the capability within them, perhaps buried or ignored, but there nonetheless. "And under the right circumstances—self-reflection, conscious personal development, a mentor, a great teacher, loving parents, a significant life experience . . . they begin to develop. Some of the leaders in our study had significant life experiences that might have sparked or furthered their maturation."[50]

Matthew ticked several of those "right circumstances"—self-reflection, conscious personal development, loving parents, and, most glaringly, a significant life experience.

By 2008, Matthew was attempting to figure out his life. To right himself to take on his next chapters. In February, he reunited with Lindsey and took her to a formal sorority dance. After she graduated with a degree in management from Indiana University in May, he proposed. They were married the following year.

Lindsey was hired as the accounting and finance manager for JL Case IH (International Harvester) Equipment in Poseyville. "I worked for a wonderful family led by the second generation," Lindsey recalled. "It was great because I got to learn from a female in a leadership position in an industry not typically run by a female. So how providential is that, that I would be entering this manufacturing industry [NIX] where there are very few women in leadership? It was great to learn from her. To stand your ground and own being a woman in this industry."[51]

While Lindsey earned a business degree from one of the most prestigious business schools in the country, Matthew's "business degree" comprised hands-on and immersive lessons in risk-taking as he ran the Outdoor Connections. Sonny watched his grandson's business ventures, and one day he approached Matthew.

"I will never forget what he said and how he said it," Matthew ruminated. As grandfather and grandson stood face-to-face in front of the cement-block building on Endicott, the shop Sonny and his father, Carl, built, Sonny poked his grandson in the chest, punctuating each word with his index finger, each a stab at Matthew's heart.

"Well, when you go broke, don't you come crawling back to me to bail you out!"

Those who knew Sonny Nix probably cannot imagine him saying such unsupportive words. People thought of him as smiling, easygoing Sonny with a heart of gold. That's one reason why his words hurt Matthew more than if they had come from someone else.

"It was so out of character for him," Matthew said. "I still don't know why in that moment he felt the pressure to say that, but I suspect he was concerned that I might crash and burn and take down the five-generation family business legacy.

"I went on to open a store, which was a massive failure, and I eventually sold off the inventory at a loss after a few years. I also ventured into building websites and video-marketing production for outfitters with a couple partners. That started to take off, but we failed because we lacked the ability to execute since we all had day jobs. All of those were tremendous learning opportunities."[52]

Matthew lost around $20,000 dollars on those ventures but has no regrets. He considers the financial loss to be the cost of his indispensable education working in the trenches.

About one year after Maureen died, Matthew began trying to convince his father to sell him half of Nix Welding. Soon after, he would vie to make his first acquisition, and Bill would hold Matthew's feet to the fire.

## CHAPTER 10

# You Just Know What You Know

Matthew had been working full-time in the business for four years when he convinced Bill to sell him half of the company in 2008. "I was a full owner at that time," Bill said. "And I thought, *If I don't do with it what he wants to do, then he's gonna quit, and I'm gonna be stuck with all of it. It's gonna go sour. Everything's gonna fold up. Why not just let him if he wants to come in there?* So he did, and it all worked out pretty good."[53]

In fairness to his three siblings, Matthew bought the business at market value. "Dad financed the sale to me, with interest, over ten years to make it more affordable for a twenty-three-year-old with no money," Matthew said. "I was ecstatic to be an owner, and it accelerated my drive to grow the business. The thought that I would be giving my dad half of that growth one day when I, presumably, would buy the other share never entered my mind."

Buying into the company created a good foundation for an entrepreneur like Matthew. "All I had to do was show up and bust my ass every day for the next thirty to forty years, and I could make a great living. It wasn't a bad deal for Dad, either. Selling me half the business then secured the fact that later he would have a buyer for the other half. It was a much better option than one day selling the assets at auction."

Although Bill never pushed his son to take over the business, Matthew was motivated to continue the family legacy of ownership into the fifth generation of Nix men.

One year after buying in, Matthew had an idea that would change the course of the shop. He stood next to Bill in the north end of their blue pole barn, pointed across West Fletchall, and said, "I want to buy that property and expand. I'll hire four or five guys and sell enough work to keep them all busy."

"You will never sell enough work to do that," Bill declared. "There's not enough work out there."

When Bill said "out there," his mental geography did not extend much beyond Poseyville. Matthew, on the other hand, saw "out there" in terms of the tristate area—Indiana, Kentucky, and Illinois.

Bill's blessing did not come easy. He was a blue-collar business owner whose accountant influenced major capital decisions. Bill and Sonny never took out a loan to run Nix Welding. They "never borrowed a dime." They paid for everything outright.

One of Matthew and Bill's trusted advisers was Tracy Ripple, brother-in-law of Matthew's mother and later a member of Nix Companies' board of directors. Tracy was in Bill's camp, cautioning Matthew against the acquisition. They and the accountant met with Matthew ("Talking down to me as if I was some young dummy, full of blind ambition"). Matthew described the meeting:

> I had already run the numbers and had scenarios to present, rebuttals prepared for each possible point made, along with spreadsheets with multiple ROI calculations and "what if" scenarios. Business financial literacy always came natural to me. I taught myself how to read profit and loss statements and balance sheets. I had a strong understanding of basic corporate finance principles. I was not making the decision blindly and was well prepared, but even so, I felt like they were ganging up on me, and I left the meeting nearly in tears. I was fighting the outside world to grow our dinosaur of a business and felt like I had to drag a giant parachute of internal pull along with me.

As difficult as that evening was, Matthew learned something valuable about himself that gave him confidence for what lay ahead. He discovered that he had as good a grasp of the company's financials as anyone in the room—maybe even better. He did not tell anyone or boast about the insight he garnered that night. He simply filed the awareness away for future transactions. The realization buoyed him.

"Sometimes, you just know what you know, and that can be liberating," Matthew explained. "I realized that night that I no longer needed to fear the unknown. To this day, I'm a huge fan of seeking outside counsel on major decisions, but I learned at that meeting to be careful about looking to others for financial guidance, no matter how 'qualified' they appear, especially if they don't have the situational awareness necessary to give good advice."

Matthew persisted with his vision, and eventually Bill consented to cosign on a loan, but he made it clear that he would not do the footwork. "Fine, if you want to buy it, I'll back you," Bill told him, "but you are going to go make the deal, you are going to get the loan, and you are going to make it work."

Matthew made a deal with the seller and went to Community State Bank, a small local Poseyville institution, to ask for a business loan. He recalled:

> Most bankers would have laughed at a twenty-four-year-old walking in and asking for $150,000 to buy a shop with no formal business plan, no equipment, no employees, and, most importantly, no cash flow to support it. Fortunately, the banker knew me and our family. He asked me if I had negotiated a price. (I had.) He asked me what interest rate I was thinking. I had looked up the prime rate on the internet just before my appointment and shot him that rate. He said that would work, and a few days later we signed the papers. No appraisal requested.

Matthew's first real-estate acquisition—129 West Fletchall, the property immediately north of the shop where Sonny, Bill, Caroline, and Matthew worked at the time—would become a catalyst for the company's eventual rocket-ship evolution.

The first real estate acquisition made by Matthew Nix in 2012 at 129 Fletchall before it was remodeled.

In the following years, Matthew would experience many more transactions with this small local bank. He became friendly with the bank president, who shared Matthew's love of flying, even taking Matthew on flights in his private plane. He helped with charters when Matthew needed to be somewhere in a hurry. Nix Companies eventually outgrew the local bank as a primary lending institution, but Matthew still values the strong relationship that gave him his start, all because of the easy familiarity established by his grandfather and father.

"Today, not only do we know there was enough work out there to keep four or five guys busy," Matthew said, "we know there is enough to keep many times more team members busy. With each passing year, we continue to chase down that work and increase the boundaries of 'out there.'"

# CHAPTER 11

# A 100-Year-Old Start-up

The more Matthew continued to develop the company, the more pressure it put on his father and the more uncomfortable it made his grandfather. "Dad and I battled constantly," Matthew said. "We had yelling matches in the shop, sometimes in front of customers, and would not speak to one another for the rest of the day."

Bill was not a fan of change, so the butting of heads was a natural outcome of Matthew's entrepreneurial spirit. The more success the younger Nix accomplished, the more Bill tried to hold his son back. "Keeping a young, ambitious guy like me humbled and grounded wasn't a bad thing; however, Dad took it to the extreme," Matthew said. "Sometimes his comments were so bad that customers and other outsiders took notice and apologized to me on his behalf."

Matthew often felt beaten down, and on at least one occasion, he told Lindsey that he wanted to quit and go out on his own. "Looking back, I probably could have made it on my own, but it would have been much tougher, and the road to get where we are today would have been much longer," he said. "I'm thankful I stuck with it, because it's so much better to build a business and succeed together."

It took many years, but Bill would finally come around and show pride in words and tone for what Matthew, Adam, and Lindsey have done to build NIX.

Every business goes through a life cycle. In his ongoing quest to learn and educate himself on how to expand, Matthew read about the philosophies of Les McKeown, founder of an organization called Predictable Success. In his book of the same name, McKeown describes the seven stages of business growth: early struggle, fun, whitewater, predictable success, treadmill, big rut, and death rattle. Looking back, Matthew said:

> Our business is unique in that it was around for more than one hundred years when I joined it. It was stable; however, on the surface, the business appeared to be in the "predictable success" stage but was actually in "treadmill." Not because the business was poorly managed but simply because the world around it was changing and it had remained stagnant. With the solid foundation that had been built and maintained by my ancestors, the business could have remained in that state for many more years, possibly even a couple of decades.
>
> My father easily could have finished out his career in that state, but by the time I joined the business, not yet knowing of McKeown's life-cycle concept, I intuitively understood that I was going to need to change the trajectory of the business if I wanted to enjoy the same standard of living that my predecessors had enjoyed.

During the exhausting treadmill stage, Matthew described the company as a 100-year-old start-up. While he takes his hat off to the previous generations, who gave him "one hell of a platform to start with," his early years with Nix Welding felt like the proverbial roller coaster ride. "Each hill we painstakingly climbed was a different challenge or misstep that had to be overcome," Matthew said. "We changed the business so dramatically that it moved *backward* on the predictable success life cycle, from 'treadmill' to 'early struggle.'"

Matthew admitted that one of his early missteps was his focus on quantity rather than quality of projects. His upbringing in the shop, chiefly Sonny's influence, had taught him to be satisfied with "good enough" repair work. That early teaching, his hunger to propagate the business, and his non-detail-oriented personality created an unhealthy combination.

He concentrated on new fabrication work as the company's growth vehicle and uncovered an opportunity to create roadside trailers that housed electronic speed-monitoring boards for the Indiana and Kentucky Departments of Transportation.

"The contract holder was a powder coater who subcontracted the fabrication," Matthew said. "After some selling, I was given the opportunity to create a prototype for them and won the bid to take over the fabrication. This was a great opportunity for us. However, the tolerances and cosmetics were more critical than I was accustomed to. Dad was still the principal owner of the company, and this contract had the potential to make him a lot of money, too. But he was not involved. I sold the job, produced the prototype, and was accountable to the buyer when they came to review it."

Several quality issues were found on the day of the inspection. Matthew quickly made corrections and shipped the finished prototype to the contract holder, but all for naught. "I had lost their confidence and was not awarded the contract," Matthew said.

He was crushed. And he knew he had to take total ownership of the process and bear the weight of the loss. "Since it wasn't anything Dad would have had to begin with, I got the sense he never felt any personal loss from the opportunity," Matthew said. "His lack of emotional connection to the project likely made his ability to coach me and point out the lessons learned much more effective. He certainly didn't miss the opportunity to do that."

An early project that gave Matthew more ideas for progress came about when a gentleman approached Nix Welding to fabricate an aboveground storm shelter from quarter-inch metal plate via rough plans he found online. During the process of building the storm shelter, Matthew got the idea that Nix Welding could build more and sell them.

"This was my first venture into building anything to inventory,"

Matthew said. "After some mediocre success at retail-selling our storm shelters locally through word of mouth and social media, I decided that focusing on the manufacturing and allowing others to do the sales was more advantageous."

With tornadoes always a spring and summertime threat in and around Poseyville, the aboveground storm shelter market was gaining traction in the Midwest. Matthew called the largest regional distributor of storm shelters and suggested that he buy from Nix Welding because, as a local company, they could save him the freight expense. Additionally, Nix Welding and the regional distributor would then stop competing against each other. The distributor "took the bait" and said he would pay Nix Welding a visit soon. Unfortunately, he showed up unannounced when Matthew was out on a service call.

"He went into the shop and saw some partially completed units and a less than inspiring manufacturing operation," Matthew recalled. "Without being there to smooth things over or present a nice finished product like I had planned, the deal was off. He continued to buy from his out-of-town source and compete with us."

The storm shelter episode was another lesson in what happens with overpromising and underdelivering. It was a valuable early experience that today helps Matthew be receptive when a team member is apprehensive about taking on a project that is outside their comfort zone.

New fabrication projects weren't the only thing that Matthew and Adam took on. Just like any start-up business trying to survive, they did whatever it took. "Early on, the first winter Adam joined the business full-time, in order to have enough work to keep us all busy, we took a multiweek job at a large hog farm," Matthew recalled. "They negotiated a lower-than-usual rate with us, and we gladly accepted it since it was an extended period of guaranteed work."

For several weeks, Adam and Matthew drove to Griffin—about fifteen minutes from the shop—to a large hog farm, where they welded and repaired metal hog pens. They entered the barn, and immediately the stench of hog manure burned their noses and stung their eyes. They worked as a team amid the horrendous odor. One moved hogs from pen to pen while the other made repairs. Often the pigs were uncooperative, and the brothers dealt with many a swine altercation. Worse, the floor of the barn was made of concrete slats to allow the hog manure to fall through to a pit below; when their welding leads ran parallel to the slats, the leads often fell through and drooped into the manure pit.

"Then, for the rest of the day, we were dragging welding leads around that were covered in pig shit!" Matthew said. "The smell was almost impossible to get rid of, no matter how hard we tried. The welding truck smelled for weeks after. We changed clothes and boots immediately when we exited the barn, but it didn't matter. The stench was in our hair and seemed to seep into our pores. Lindsey must have, again, wondered if my big ideas would ever amount to anything when I came home smelling like hog manure."

Today the brothers still have their share of challenges, but looking back on jobs like the hog barn escapade keeps it all in perspective. "I won't ever ask our team to do anything I have not done or would not do," Matthew said. "They get into some less than desirable jobs at times, and I certainly appreciate their tenacity, but they aren't going to get too much sympathy from a couple of guys who drug welding leads through hog shit for weeks on end just to pay the bills.

"Grandpa Sonny said, 'I never hated going to work. There were some jobs I didn't love, but I always loved my job.' That sums it up. If you really love what you do, and you are driven toward a goal or a mission, you don't let a 'shitty' job get you down or steal your belief in the dream. Things can change in a hurry."

No truer words have been said. Next up: "Build a yacht, and they will come."

**CHAPTER 12**

# Rite of Passage

In 2008, New York real-estate man Joe "JZ" Morris, a tall, broad-shouldered fellow with shorn white hair, sauntered into the blue metal pole barn of Nix Welding and asked Matthew if he would like to build a seaworthy custom metal yacht.

Matthew had met Joe Morris one year earlier when the New Harmony, Indiana, native retired and moved back from New York City, where he had spent the bulk of his career in commercial real estate. Joe's late father was a well-known financier from Southern Illinois with offices in New York and Florida who had owned numerous grain elevators in the tristate. He began the predecessor company to Berry Plastics in Evansville and had done business with Sonny.[54]

"When JZ moved back and needed some metalwork done, he came to us based on our generational business relationship," Matthew explained. "When he approached me about his boat idea, the biggest projects we had done for him to that point were a custom stainless steel range hood for his kitchen and a wrought iron fence for his yard. Neither qualified us to build a high-end custom yacht that would take two years to build. But that didn't scare him."

"I had been using the company a lot of my life," JZ was quoted as saying in *Evansville Business* magazine. "There was no other option; they were my choice, and they were first-rate throughout the two-and-a-half-year process."[55]

That process was a deep and wide learning experience for twenty-four-year-old Matthew, who faced ridicule for taking on such an extraordinary, massive venture in his family's welding shop. This was not a typical farm-implement job that pulled up to the garage door for a quick weld by Sonny or Bill.

JZ sailed the world. Coastal sailing, he clarified—not ocean crossing. He had studied naval architecture at the esteemed Westlawn Institute of Marine Technology, School of Yacht Design, based in Eastport, Maine, and had been a US Coast Guard–licensed master captain for twenty-five years, owning a number of boats.

JZ wanted Matthew to craft a fifty-foot steel-hulled yacht narrow enough to traverse rivers yet seaworthy. The yacht would be named *Passage*. Appropriate, because for Matthew, the endeavor would be a rite of passage.

"When he asked me about the project, naturally I was all in," Matthew recalled. "I had been waiting for an opportunity to grow the business for years, I loved boating, and this project was more of a work of art than it was a precise fabrication. All of those aspects made the request right up my alley. Grandpa liked to say I was 'just young and dumb enough to try it.' The older I get, the more I know he was right. Thankfully, I *was* young and dumb enough to try it, because it was the foundational project that gave me so much confidence for the future."

However, Matthew did not agree to build the yacht right away. After all, Nix Welding didn't have a space large enough to build a fifty-foot boat. He told JZ he was interested and they could talk further soon. About two weeks later, JZ phoned.

"Do you have a passport?" he asked.

"No, I don't."

"Well, get one coming. Get it expedited. We are going to England to meet my architect."

As great as that sounded, Matthew wisely hesitated. "Wait a minute. I haven't committed to anything."

"That's okay. You don't have to commit to anything," JZ said. "I just want you to come along and represent me from a fabricator standpoint."

Matthew and Lindsey were newlyweds when JZ made his boat-building proposal. She remembered, "Matthew came home from work that day and said, 'JZ wants me to build a yacht, and he is going to take me to England. I gotta get a passport.' I said, 'How are you going to build it?' He said, 'I don't know.'"[56]

So, Matthew obtained his passport and headed for an overnight in New York City, then on to the UK as a "fabrication consultant." His title would soon change to "yacht builder." Matthew had never been to New York or overseas.

Joe "JZ" Morris and Matthew Nix relaxing after a day of visiting Branson Boat Design in Peterborough, England, in 2008.

In England, JZ and Matthew looked at boats and met naval architect Nick Branson of Branson Boat Design LTD, a company renowned for its modified Dutch barge designs.[57] "Once we met the architect, despite the thick English accent and strange words, I found him to be very down to earth and knowledgeable about the fabrication/build process," Matthew said. "He wasn't a stereotypical architect who knows how to make it look nice on paper but has no idea how to make it actually come to life. He had experience building one of these boats, and talking to him put me at ease."

The first night after the meeting, Matthew called his father from England and said, "Well, we are going to build a metal yacht."

Matthew had not yet told JZ he was all in, but in his heart, he knew he was. He didn't have a building to house the work at that point, but that was a minor detail he would figure out later.

When he arrived back in the States, he began looking for a building site. He wanted to stay in or around Poseyville for the ease of moving materials between the two shops. He considered a few options, but they were too far from the primary shop. He decided to approach the owner of a building that sat kitty-corner from the welding shop on Fletchall and ask if he would be interested in renting out space. (The facility later became the main office, engineering office, and custom fabrication shop for Nix Companies.)

"To my surprise he was willing to move his inventory to make space for me at a very reasonable rent rate," Matthew said. "The lesson here was never assume anything in business or try to think for the other party. As I like to say now, 'It's free to ask,' and I try to instill this way of thinking with my team."

With the shop set up across the street from Nix Welding, Matthew was officially in the yacht-building business. The first step was to have the metal CNC-cut (computer numerical control) using architect Nick Branson's computerized "cutting files." CNC plasma cutting is the process of cutting metals using a plasma torch controlled by a computer rather than a hand torch. CNC cutting provides maximum precision. JZ had started that process with another shop in Mount Vernon, but when the project did not progress to his liking, he allowed Matthew to take over.

"I was so uneducated about these things at the time that we went all the way to Owensboro, Kentucky, to find a shop to plasma-cut our plate material," Matthew said. "Had I known better, we could have simply had our steel supplier cut the plate and ship it straight to us complete. My guess is I found the place by Googling 'CNC plasma cutting.'"

In an *Evansville Courier & Press* story, Matthew was quoted: "Bending steel is hard work, and the boat is all about curves. There are very few straight lines on this vessel, even in the detail work."[58]

Matthew's father, Bill, looks back on the project with a tinge of amazement. "Some of the stuff Matthew did I wouldn't even attempt. Like the fifty-foot yacht that he built. Bending all that steel cold, pulling it, stretching it, welding it. It was like, *Boy, you can't bend material like that*. But he did."

The plasma-cut parts started to arrive, and the excitement began. Matthew's brother was on summer break from the University of Southern Indiana and was working at the shop. Adam began to lay out the build frames that would support the boat during its two-year construction process.

"Fortunately, I had learned many quality lessons by then and knew that if this part of the project wasn't done meticulously, the rest would be a nightmare," Matthew said. "We used a transit to shoot grade, then shimmed and leveled for days, dealing with an old, unlevel concrete floor."

"Shoot grade" refers to finding how to level a section of land relative to the surrounding area. Once the build stands were anchored in place, the keel was set. This is the symbolic first step in any boat-building process. Setting the keel when building a boat is equivalent to breaking a bottle of champagne over the bow at the completion of the build.

Next, the hull plates were put in place, and the hull started to take shape. Then the framing went in. This was the skeleton of the vessel, giving it form and strength during the build. Once complete and tied into the hull, the framing, surprisingly lightweight (relatively speaking), formed a strong structural channel. The curves of the hull also provided strength.

The yacht Matthew Nix built for JZ Morris under construction, 2009.

"All of the frames are CNC-cut to shape, and all of the hull plate is wrapped around those frames, pulled into shape with chains, ratchet boomers, come-alongs, dog screws, and a four-pound hammer," Matthew explained. "I've often joked about the primitive and fairly few tools one would need to build a yacht."

At the time, Lindsey worked as an accounting and finance manager at JL Equipment in Poseyville. "I would go down every day after work to see the progress on the yacht and how it was taking shape," Lindsey recalled. "The building was basically falling down around it. That's how old the building was."[59]

As the yacht-building process heated up, so did the stress of managing the regular workload in the welding shop. Adam returned to school in the fall, and without his help, the tension in the shop picked up even more. "When I spent my days working on the boat, Dad was irritated I wasn't helping him," Matthew said. "When I was helping

Dad and Grandpa with the normal farm repairs, JZ was irritated that I wasn't staying on his build. It was a no-win situation, and I was taking the brunt of it."

In the middle of the yacht build, on a hot and humid day in June 2009, Matthew and Bill were standing in the bathroom of the shop, washing the dirt and grease off their hands like they did each day before they left for their lunch break. Sonny, who by then worked three days a week, approached.

"Call Aramark and have them cancel my uniforms," he told them. "I'm done."

"Just like that, at noon on a Wednesday, Grandpa hung it up," Matthew remembered.

Sonny turned seventy-five on June 15. After his fifty-two years of welding and managing a shop, the abrupt, understated manner in which he delivered his retirement announcement is now legendary.

Throughout the yacht project, Matthew's nightly dreams were filled with details of "building a boat in [his] sleep and waking up fifty times with measurements and calculations racing through [his] head."[60] Plus, he was still running his entrepreneurial endeavors: the Outdoor Connections, with its retail outlet, and his marketing ventures. All would soon have to be shuttered and chalked up to a hands-on education.

Matthew needed help. Serendipitously, a teenager who was enrolled in the high school welding program and lived down the street showed up looking for a job, eager to work and learn. The young man lived with his grandfather, a truck driver, and the two had been in the shop previously for trailer repairs. Because no one outside the family had ever been hired at the shop, Matthew approached his father with trepidation about the young man.

"Dad, he's here wanting a job. He wants to be a welder."

"Well, hire his ass then!" Bill barked.

Donald Yancy worked after school and on Saturdays. When he graduated in May 2011, he was hired full-time and has the distinction of becoming Nix Companies' first full-time nonfamily-member employee, bringing the total number of Nix Welding employees to five. (In 2024, the number of team members will have grown to about 200.)

He worked alongside Bill and Adam for around ten years, but when the business grew, he left, admitting he liked the shop better when it was smaller. "It wasn't a good fit anymore for either party," Matthew said. "That's an unfortunate part of growth and change. Some people come along for the ride, and some don't."[61]

Soon, word was out about a "yacht being built in the middle of cornfields," and the shop would become a revolving door for tours.

## CHAPTER 13

# Yacht in a Cornfield

So many people came by to see the yacht during construction that the running joke was Matthew should charge an admission fee. "It was sometimes a bit of a distraction," he admitted. "But the publicity that it brought to our business was ultimately a great thing." Nix Welding and *Passage* were featured in *Evansville Courier & Press*, *Posey County News*, and *Evansville Business* magazine, as well as on three Evansville TV stations. Before the yacht, few knew of Nix Welding. *Passage* put NIX on the map.

"Family was always in to see it, and everybody was driving by," Lindsey said. "It was probably the talk. I'd go to lunch with my coworkers, and they would ask, 'Can we go see it?' I think a lot of people thought Matthew was crazy for doing it."[62]

Matthew prized the yacht for the publicity it reaped and the self-assurance it fostered in him. He said:

> I often tell people that the money we made from this project was the least important thing. The bigger value came from the notoriety Nix Welding gained and the confidence it gave me to spread my wings and grow the business. In the years to follow, I can't tell you how many sales calls I would go on and people would say, "Oh, you are the ones who built that yacht." When a customer had an unusual or large project, I would try to casually work the yacht into the conversation. At that point,

it was the most prominent project in our marketing material. Despite its little relevance to the commercial and industrial fabrication work I was trying to sell, customers often said, "Well, if you can build a yacht, you can surely build this."

JZ described *Passage* as a steel-framed "displacement vessel" reminiscent of European canal boats, based on a Dutch barge design. The blue-and-white steel boat was designed to float in 3.5 feet of water with a 210-horsepower Cummins diesel engine manufactured in Columbus, Indiana.

John Crum, a master carpenter and cabinetmaker out of Mount Vernon, crafted the interior with Posey County–harvested solid cherry. It was outfitted with fully appointed and air-conditioned sitting and sleeping quarters, a kitchen, and a bath. Similar to a luxury recreational vehicle, *Passage* could sleep four adults comfortably. JZ said the wood paneling in the downstairs living quarters "looks just like [his] library at home."[63]

• • •

## THE LAUNCH OF PASSAGE

After two years of hard work, in April 2010, the long-awaited launch day arrived. At 5:30 a.m. the day prior, Matthew and his crew started the preparations. A crane was delivered. At daylight, *Passage* was rolled out of the shop. After hours of meticulous maneuvering, it finally sat securely on a stretch lowboy trailer. JZ paid an off-duty police officer to guard the boat all night. With all the PR exposure, the fear was that the yacht could be vandalized.

The next day, *Passage* began the twenty-five-mile trip from Nix Welding to the Ohio River at Port of Indiana-Mount Vernon. The Port of Indiana-Mount Vernon handles more cargo than any other port in the state. With close proximity to the confluence of the Ohio and

Mississippi rivers, it serves as a major gateway for agricultural, energy, and manufacturing industries.[64]

From the get-go, the trip was stressful. In the shop parking lot, the boat became "high centered," with a section stuck on higher ground than the wheels.

"We had to bring in a four-wheel-drive tractor to drag the truck and trailer out of the parking lot onto the road," Matthew recalled. "The trailer had to be stretched out so long to accommodate the boat, so it only had about two to three inches of ground clearance. Even the crown in the road could be too much for it to clear."

Lindsey, Donna, and Matthew's sister Lauren hopped into the car and followed the boat to the port down Southern Indiana's winding byways. "We were all excited," Matthew's mother remembered. "The road down was not straight. There were some turns. It was just so cool watching that big boat on the highway. It was crazy."[65]

"Finally, after a very stressful couple of hours, we arrived at the port," Matthew said. "But the excitement was far from over."

The 50,000-pound yacht with a showroom finish had to be lifted off the trailer, boomed out over the river, and lowered about thirty feet alongside a concrete wall into the rolling, muddy Ohio River.

Matthew was one of four people permitted to ride the boat down while it was hooked to the crane. The other three were JZ, owner and captain; John Crum, master woodworker and builder of the yacht's interior; and Bradley Stevens, a local welder who helped Matthew with the build.

"As a crane put it in the water," Donna said, "all I could think was *I hope as heavy as that is, made of metal and iron, that it will float.* It was a bit lopsided. I talked to Matthew about it later. He said, 'Oh yeah, we just had to adjust the weights a little bit.'"[66]

Lindsey had the same worry. "I remember thinking, *Is this thing gonna float?* Because you have no idea until you put it in the water."[67]

Sonny, Bill, and Caroline remained at the shop that day. Duty called in two directions, and each Nix chose their course.

Matthew described riding the boat down into the river. "As the boat touched the water and the crane continued to cable down, it slowly sat deeper and deeper in the water. Once in, but prior to the crane letting us go, we fired the diesel engines to allow them to come up to temperature."

*Passage* on the crane that lowered it into the Ohio River.

As the men hovered in the boat with its engine firing, still attached to the crane, JZ checked the gauges. Suddenly, he called out, "We've got water in the bilge!"

Matthew's heart sank, and they both frantically broke into action, checking every point of entry. "All valves were checked, and although I had meticulously inspected all the welds with die penetrant prior to paint, I began to look for leaks on the bottom of the hull," Matthew remembered. "Finally, we discovered the culprit in the engine room. The seal on the main drive shaft that goes from the transmission to the propeller was dripping."

Fortunately, there was never enough water accumulation to put the vessel in jeopardy. After a few minutes that felt like an hour, the issue was solved, and they radioed the crane operator to continue the boat's descent. As the yacht lowered fully into the water, the operator, over his radio, counted down the weights from the scale on his boom: "Twenty thousand, fifteen thousand, ten thousand, five thousand . . . You've got it all!"

Matthew will never forget those words. The scale on the crane boom read "ZERO."

*Passage* was afloat.

*Passage* meets the mighty Ohio River.

The men released the shackles from the slings, dropped them overboard, and began the short cruise a half mile downriver to Mount Vernon Barge Company, where *Passage* harbored for a few days for some finishing touches. She looked out of place, a gleaming diamond in the murky river next to the rusty docks of the barge mooring. For Matthew, the sight of her there, the juxtaposition, symbolized the journey his family's business had taken from blacksmithing to farm-equipment repair to yacht building.

"For months after launch day, I continued to wake up in the middle of the night from nightmares surrounding some type of catastrophic failure," Matthew said. "Whether it was a cracked weld, poor weight distribution, or a number of other possible factors."

At first, *Passage* reposed at Evansville's Inland Marina, where JZ offered it for sale for slightly less than $1 million.[68] Later, JZ navigated *Passage* to the Tennessee River, then Kentucky Lake, and headed down a canal system into the Gulf of Mexico to spend his summer yachting the waters near South Florida.[69] Matthew went with JZ on the leg to the Sunshine State, from Evansville to Kentucky Lake. Matthew's father, Bill, went along on the second leg from Kentucky Lake to Pickwick Lake in Alabama.

A story about *Passage* in *Evansville Business* magazine spoke to Matthew's love of his native town, its people, his family, and his craft:

> Because yachts are supposed to have curves and "nice yachts shouldn't have straight lines," says [Matthew] Nix, *Passage* has "really beautiful curves." These beautiful curves are built at the beautiful bend in the river, Southwest Indiana, where [JZ] Morris and Nix call home, which is why two landlocked men created a local boat company here. "I love to travel, but this is home," says Nix. "If I'm going to run a business and have a family, I want it to be here." Near the Ohio River, where boats and yachts have access to anywhere in the world, Nix says.
>
> "Also, we have an abundance of good people in our area," says Nix. "To build a custom yacht like ours at a reasonable price, it takes people who know how to give an honest day's work and have pride in what they're doing."[70]

These words from Matthew's youthful twenties predicted the Nix Companies' core values of today. A few months later, Lindsey left JL Equipment to work at Nix Welding. As she made that transition, she and Matthew took a weekend getaway to Sarasota and stayed with JZ. They boarded *Passage* with JZ at the helm. An American flag unfurled at the stern. "It was so cool to be on it," Lindsey said. "We went in the little intercoastal waterways, and the dolphins just loved the hum of the boat. They swam alongside."[71]

A dolphin hitches a ride in a boat's wake to be happily pulled along by the wave. Like the dolphins, Matthew would continue to ride the swell created by *Passage*.

*Passage* afloat in Florida.

# CHAPTER 14

# Boat-Building Blues

In the midst of building *Passage*, an elderly oil man from Southern Illinois, Richard Hoover, heard about the vessel and contacted Matthew about building a custom metal yacht to traverse the Ohio River. Matthew would soon learn that both the project and its owner were very different from *Passage* and JZ.

"He was what we call a 'good ol' boy,'" Matthew said. "Little education, roughneck, self-made millionaire. The story has it, he saved up money to drill his own wells. He drilled two or three dry holes and had enough money left for one more. He hit oil, and the rest is history."[72]

Hoover continued to drill more and buy more oil rights. He owned more than a hundred rental houses when Matthew and the NIX team built his yacht. "I respected the hell out of how he came from nothing, but that was where my love for him ended," Matthew said.

Despite his millions, Hoover wanted a high-end, seventy-foot custom yacht created on the cheap. He continually changed his mind on what he wanted, and he did not want to spend money on proper engineering or a marine architect. "After much convincing from him, I ended up designing the yacht myself," Matthew said. "The framed drawing in our conference room is the final version I drew at my kitchen table. I love architecture but was a little skittish when it came to this. As Dad and Grandpa said, I was just young and dumb enough to do it."

Hoover wanted the vessel to resemble a tugboat like those seen on the Ohio River. The boat was roomier and more practical than *Passage* but was not as pretty (or "sexy," as some say in the boating world). It had two large diesel engines. The custom wood interior was made by John Crum and his crew, who did the cabin for *Passage*. Matthew fabricated a custom-built interior spiral staircase from the main cabin up to the pilothouse. The boat was made of steel, but the pilothouse was aluminum, which created some engineering challenges. This boat was Nix Welding's biggest project to that point.

"I used common sense and consulted Dad and Grandpa," Matthew said. "Sometimes we don't need to overcomplicate things. It was so tall that the pilothouse had to be hauled to the river on a separate semitrailer, and we never did a test fit because our shop wasn't tall enough. The wiring for electronics and hydraulics for steering had to plug and play once the crane placed the pilothouse at the river."

The biggest challenge when engineering a boat is weight displacement, involving complex calculations. Changes made can compound depending on many variables—how far aloft, how far outboard, how far forward/aft, etc.

"In the end, I used my gut feeling and some advice from JZ," Matthew recalled. "At the very last second of the build process, I moved the pilothouse forward because I was skeptical we were going to be too heavy on the stern. I was a nervous wreck on launch day. I had nightmares about it for months afterward—maybe years. But in the end, it floated perfectly on its lines. Dumb luck, maybe."

Hoover was happy with his boat, which he docked at Golconda, Illinois. He named it for his sons and grandsons—*Hoover's Boys*.

After the second boat project, Matthew knew boat building was not part of his vision for Nix Welding. *Hoover's Boys* had done a number on him. However, he saw the broadening scope of what the company did as the impetus to change the DBA. While today the official name of the primary operating business is still Carl A. Nix Welding Service Inc., in 2010 Matthew changed the DBA from Nix Welding to Nix

Metals. He knew more metal fabrication was on the horizon, and he wanted the name to reflect that.

Lindsey's father, Don Tenbarge, today a retired UPS driver who worked part-time for what became Nix Companies, watched Matthew grow from a fifteen-year-old teen who dated his daughter into a young entrepreneur and then a seasoned innovator. "I think, for the most part, the boats were a phase of him growing into this business," he said. "Building boats was not going to be his future. . . . It was a learning experience, a stepping stone to where he is today. Matthew had the vision to know that metal fabrication was the key to the future of where he wanted to go."[73]

## CHAPTER 15

# Doubling the Business Overnight

Matthew encouraged his brother to join the business. The two talked about it during Adam's senior year at the University of Southern Indiana. At the time, there were four full-time Nix Metals team members: Bill, Matthew, Caroline, and Donald Yancy. Matthew revealed his entrepreneurial aspirations to Adam, overlooking his brother's deep aversion to the dirty, backbreaking welding shop that their mother once used as a last-ditch punishment for Adam's bad behavior.

"I hated the welding shop," Adam admitted, looking back as the vice president of operations and principal partner today. "I mean, with a passion. I hated it. I can't underline that enough. I worked there in the summer, so it was always hot. I played baseball, so I felt like I spent my fair share of sweating and putting in the work outside of the shop. Because when you go to the shop, you're gonna clean."

Adam's decision to take Matthew up on his offer required a leap in perspective. What made the difference was that Adam was fully on board with his brother's ambitions for evolution. Where Matthew is a "driver," Adam is an analyzer, holding back while he thinks about next steps and speaking up when the time is right. Their different personalities make for a good business balance.

"I still wasn't crazy about joining the business," Adam said. "But I didn't have many options, and Matthew was enthusiastic about the possibilities, so I figured why not? Let's give it a shot."

Adam received his degree in marketing on May 8, 2011. "I

graduated on a Sunday, and I started work on Monday," he continued. "I asked Dad if I could have a few days off before I started, and he said, 'You can start tomorrow morning.' We went to Hacienda [a restaurant] for dinner right after graduation. Big celebration," Adam added with sarcasm. "I was with my whole family—Dad, Mom, Matthew, Lindsey, Jennifer, Lauren, and my girlfriend Lacey, who is now my wife. We got home at seven o'clock at night, and I went to work at Nix Metals the next morning at seven thirty."

A couple of weeks later, a second nonfamily employee joined the team after graduating from high school.

"With myself and one other new team member coming into the shop, backfilling Matthew, it gave Matthew an opportunity to grow the fabrication portion of our business," Adam said. "We now had two more people on the payroll full-time and had no intentions of scaling back to where we were prior."

Adam admits that he wears his emotions on his sleeves most days, and others take to that. He had watched his father rule with an iron fist and knew that style did not work for him.

"Dad exhibited tough love," Adam said. "I really care about how people feel when you interact with them. Customers come in with a problem, so they're not happy. They're stressed, and it's my job to help them. That helping approach was the way I went about my business and handled customers. Grandpa Sonny was like that. He always enjoyed dealing with people. He felt blessed that people came into the shop and wanted us to do their work. I took that from him."

However, Adam has always admired his father's work ethic. "When I began working full-time in 2011, I fully immersed myself in learning the trade and paid close attention to how Dad went about doing his work. I wanted to emulate him and gain the respect the customers had shown him and my grandpa over the years."

About a year after he joined full-time, Adam asked his brother for additional responsibilities, and Matthew allowed him to create their first company handbook.

The next big break for Nix Metals came in the form of electrical equipment houses—e-houses. Nix Metals had been modifying a few shipping containers for a new customer that built electrical controls and automation equipment for the mining industry. Matthew's contact was Brett Seibert, a local guy in charge of the e-houses that held the equipment.

"Modifying the containers was certainly better work than fixing a broken manure spreader," Matthew recalled. "But it wasn't the volume of work to push us to the next level. At some point along the way, Brett mentioned 'the other buildings' he commissioned companies to build. After asking some probing questions, I came to realize that the containers I was modifying were only a small percentage of the number of buildings his company utilized. Most of their equipment could not fit within the parameters of the shipping containers Nix Welding was modifying, which were standard dimensions."

Matthew asked Brett, "What do you do when you can't use containers?"

"We have e-houses built. You wouldn't be interested in quoting those, would you?"

Having just built a yacht, Matthew was not scared of much, so he asked for more information. Within a week, he and Brett made the two-hour trip to Western Kentucky to view the most recent batch of buildings Brett had ordered. Matthew took photos and did some crude reverse engineering in the field that day.

As luck would have it, Brett was preparing to place an e-house order for a new mine being built in Southern Illinois. Matthew asked if he could bid on it. Brett sent him the drawings. "The only kicker—he needed the pricing back in two weeks, and during one of those weeks, I had a vacation planned," Matthew said. "So in the first week of December, with drawings in tow, I left on a trip with Lindsey. At that time, I probably had not bid a $50,000 project, much less a $500,000 opportunity, but as they say, you tackle it like you would eat an elephant, one bite at a time. I finished the estimate on the flight back with my yellow legal pad and calculator. (I still like those two tools.)"

Upon their return, Lindsey helped Matthew type up the twelve-page quote, which included the full specifications of the building down to every detail. Over the following week, Matthew answered a series of questions, had an in-person interview with Brett's boss, and by Christmas it was evident that Nix Metals had a good shot at winning the job. Making the stakes even higher, the project had a tight timeline. The customer wanted to start taking delivery within weeks of the order, and all units had to be completed by the first of May.

"We were unproven and, frankly, inexperienced, so Brett was taking a huge leap of faith by offering us the contract," Matthew admitted. "To this day, we owe a great deal of our growth journey (or at the very least its acceleration of it) to his decision to give us that first big project. Between Christmas and New Year's Day, I got the official word that we were receiving the order, and by the first week of January 2012, I had a purchase order in hand for close to $500,000."

Matthew recalled the blood-pressure-raising significance of this deal. "To give some perspective of how important and risky this order was, the prior year Nix Metals only did about $680,000 in annual revenue. Our annual revenue in 2012 would be nearly $1,250,000; with this one order, we had almost doubled our business overnight. Now all we had to do was execute. Minor detail."

Nix Metals did not have enough equipment, the right equipment, or enough workers to perform the job. There were not enough hours in the day for the job to be completed on schedule while maintaining existing customers without at least tripling the workforce.

"Also, cash flow, the one thing that many business owners neglect, had to be addressed," Matthew said. "The complete contract might have been very profitable, but since it was nearly equal to the sum of our entire annual sales, we could not possibly finance the project until completion. At that time, our credit limit with the steel supplier was roughly $20,000 a month, and we had no line of credit with the bank. Today we can spend millions in a single month on materials.

Fortunately, Brett worked with us once again."

Matthew negotiated 40 percent down on each unit once they started building, and the remaining balance would be paid at shipment of each unit instead of waiting to be paid when the entire project was complete. This allowed the shop to stay cash-flow positive throughout the entire project, with some of the down payments used to fund necessary equipment purchases. The payment terms negotiated for this project remain the foundation for requested terms on large endeavors for Nix Industrial.

One of the needed purchases was "Big Red," the largest forklift the company owns, still a staple in the fleet. Matthew recalled, "I found Big Red on eBay and purchased it sight unseen for $7,000. Two good ol' boys from Alabama delivered it on a lowboy trailer pulled by a semitruck that looked as rough as the forklift. They didn't arrive until almost midnight due to a blown tire they had to change while on the way. I met them at the shop in the middle of a cold January night, and we unloaded it."

Nix Metals had no way to load the e-houses onto delivery trucks until Matthew figured out a way. "After calculating the max weight of the largest e-house, I assessed that we could buy one 15,000-pound lift, place it on one side of the e-house, and then put the two other 7,000-pound trucks we had on the other side of the e-house. With the three forklifts all lifting in unison, a semi could back underneath the e-house."

The $7,000 fork truck paid for itself many times over, and several years later it was completely restored at a cost of $30,000. "A new asset of this size would cost well over $100,000, and there is nothing on this machine we can't repair or rebuild, so it made perfect sense," Matthew said. "This kind of resourcefulness has allowed us to grow our company at a staggering pace without exhausting all of our working capital on fancy equipment. As we continue to grow, I want to instill this same resourcefulness in my team. Just because we are one hundred times larger now doesn't mean we always need the fanciest equipment. Sometimes it makes sense to buy new, and sometimes it's not necessary because it's wasting critical resources that could be deployed elsewhere."

Several welders, another band saw, and many other miscellaneous pieces were also purchased. Next, Matthew needed to quickly find at least six more employees. The deadline loomed. This was Matthew's first real experience with recruiting. Today, this activity is a full-time position in the organization, but at this point, it was up to Matthew. He placed ads for welders/fabricators anywhere he could and started sifting through the applications. Since Nix Metals was totally unknown and located in Poseyville, they didn't get the type of top-notch talent they have today. "Basically, if you had a driver's license, were not a violent felon, and could pass a basic weld test, you were hired," Matthew said. "It was really bad."

Within a few weeks, the workforce was complete. The next challenge was to ensure there was enough leadership to manage both the existing legacy business and the new industrial business. A plan was made. The e-houses would be built during a night shift from 3 p.m. to 1:30 a.m., four days a week, to not disrupt the existing business that occurred during the day. Adam and Matthew shared management of the night shift.

Adam was living with former college buddies thirty minutes away on the west side of Evansville, so to make the commute easier, he moved in temporarily with Matthew and Lindsey, who lived within walking distance of the shop at the time. As the project progressed, the team developed a rhythm, and Matthew and Adam were able to take a dinner break between the day and night shifts and sometimes grab a few hours of sleep before returning to the shop at 1 a.m. to lock up. The team members back then could not be trusted to lock up. "Contrast that to 2018," Matthew said, looking back. "When we acquired Superior Fabrication one hour from Poseyville, it was several months before I realized I did not even have a key to the building there."

They also worked many weekends, thus working seven days a week. "About six weeks into the project, the deadline was approaching to ship our first unit," Matthew said. "This was important for two reasons: First, this was our first milestone to maintain our customer's confidence, and second, having nearly burned through our forty percent down payment, it was imperative that we be able to invoice for the remaining balance of the first unit to maintain positive cash flow."

The delivery truck was set to arrive on Monday morning. The interior paint of the e-houses was completed the prior Saturday, hired out as a side job completed by Jason Offerman and his friend John Dike. Jason worked at a body shop in Evansville, and both men are now two of Nix Industrial's longest-tenured team members. Jason became the first employee in the coatings shop, then shop foreman, and later transferred to the Field Coatings division when it was launched, where today he is the shop coordinator. John was the second hire in the coatings shop and today is the shop foreman there. (In appreciation of their years of service and loyalty, both Jason and John were among the first ten-year team members who were treated, along with their spouses, to an all-inclusive trip to Cancun, Mexico, in 2024.)

Matthew Nix stands next to an e-house installed in a coal mine, 2012.

In May 2012, Nix Metals shipped the last e-house two weeks ahead of schedule. Annual revenues had doubled, and staff had increased by six. Much-needed equipment had been acquired, and the team had proven they could do much more than repair farm implements. Matthew had already secured another contract to build another dozen e-houses for a mine under construction in Owensville. However, he knew the e-house contracts would come to an end because there were no new mines on the horizon. He did not want to lay off the men he had just hired, so he began to think of ways to bring in more business.

"We had sorted through several guys and had a fairly decent crew," Matthew said. "I never felt it was fair when companies staffed up for big jobs and then kicked guys to the curb when the work dried up. To this day, it is a principle we hold, and it started with the e-houses."

• • •

Lindsey had left JL Equipment to join the business in 2012 when Matthew was starting the Paint & Powder Coating division; he knew someone was going to have to sell the new division, and Lindsey was a good candidate. "I thought, *Great*," Lindsey said. "*I can get back into marketing and sales.* I told my mom I was quitting my job, and she said, 'How are you going to have health insurance?' We were this small family business, so I said, 'We're going to buy it.' At that time we were spending a lot on benefits. We knew it was going to pay off one day. Matthew is this great visionary, and I thought, *He's not going to take us down the wrong path.*"[74]

Back in the middle of February, when the fabrication of the e-houses was in full swing, shop space was minimal. As soon as the exterior siding was complete on a house, it was moved outside, where the finishing was done. Heaters ran inside the e-houses to dry the paint overnight. With the tight deadline, Lindsey was pulled into service to paint. She wore many hats in the early years.

On a particular Sunday when it was bitter cold, Matthew and

Lindsey worked on stepladders inside an e-house that sat outside under Indiana's gray winter sky. As the couple slogged side by side, caulking joints and cleaning up, Lindsey good-naturedly asked, "Tell me again why we are doing this?"

"Someday it will all be worth it," Matthew replied.

"I still ask him why we are doing this," Lindsey said with a laugh in 2023. "A few weeks ago, when stuff was hitting the fan, I looked at Matthew and asked, 'What are we doing?' And he said, 'If it were easy, everyone would do it.' He tells me that all the time."[75]

Don't be fooled. Lindsey is all in. She has Matthew's back as his cohort in life and business. "Any time Matthew comes to me with a crazy idea, something 'we gotta do' or a new business venture or an acquisition, I sometimes have reservations but never any doubt he can do what he says he can do. It's going to happen. And we've got the team that's not going to fail. So most of the time, I say, 'Let's rock and roll.'"

Matthew gives Lindsey credit where credit is due. He tells people, "I may be the head of the company, but she's the neck. You don't turn a head without a neck."[76]

Lindsey's father described her this way: "She is very caring, with lots of energy. She and Matthew feed off of each other. They bounce ideas off one another and are a great match. She's very sensible and opinionated in a good way, because she has good insight."[77]

When Matthew said they needed to hire their first salesperson, Lindsey was by his side through the interview process. Today, Matthew's response as to *why* they do what they do has not changed. He continues to strive with intensity, enjoying the journey and declaring, "Someday it will all be worth it." Like hot iron on an anvil, "someday" is not hammered out until it's shaped the way the smithy envisions it.

## CHAPTER 16

# First Salesperson

Adam Schmitt joined the team as Nix Metals' first industrial sales and marketing person in the summer of 2013. Schmitt had no experience in the industry. His background was in medical staffing. So why Adam Schmitt?

The sales position was the first professional post Matthew hired, so, prior to conducting interviews, he sought advice from someone experienced on the topic, as he often does. His brother-in-law Jeff Hite, married to Lindsey's oldest sister, Kelly, was a sales manager for Cintas Corporation in Fishers, Indiana. Cintas is a Fortune 500 company headquartered in Cincinnati, Ohio, that provides businesses with uniforms, cleaning supplies, safety needs, and more.

Cintas has been named several times to *Forbes*'s list of "100 Most Trustworthy Companies in America," which recognizes transparency in accounting and governance. Honesty and integrity form the core of Cintas, a family-founded business that began in 1929. A business Matthew wanted Nix Metals to emulate. Coming full circle, today, Greg Eling, vice president of Cintas and the company's longest-tenured employee, sits on the board of advisers of Nix Companies.

Matthew and Jeff often talked shop, and early on he was intrigued by what Matthew was doing to cultivate the company. When they first met, Jeff quickly realized that Matthew's business persona resembled that of someone much older and more seasoned. "I was drawn to learning from Matthew," Jeff said. "After graduating from college, many of my friends went on to graduate school or white-collar jobs in the financial sector. I was drawn to the industrial laundry business. It wasn't the glamour of picking up dirty uniforms that appealed to me; it was the corporate culture at Cintas, similar to the environment at NIX today. I never worried about 'blue collars' or 'white collars.' I wanted a career with a company that would invest in me and my future and allow me to feel good about my work. That sort of employee investment and care is like that offered now at NIX."

Today, NIX team members wear uniforms from Cintas, but that was not the case back when Jeff and Matthew first knew one another.

Jeff graciously provided Matthew with hiring guides that Matthew, Lindsey, and Adam Nix used to interview five candidates, including Poseyville native Adam Schmitt. Jeff also had conversations with Matthew about aspects of the hiring processes that he had learned at Cintas. "Because of our relationship, I felt like I could add a little value," Jeff said in 2023. "I appreciated what they were doing—trying to grow a sales force—because I was a lifelong sales guy."

Jeff would also facilitate a tour of Cintas during a NIX executive retreat in the early years. "I got with my boss, and we thought not only could we give a tour, but we could get some of the Cintas executives in a room with NIX executives," Jeff said. That's when Matthew first met Greg Eling. Jeff saw the event as an opportunity to create a symbiotic relationship that could help both his brother-in-law and Cintas. Jeff knew that as NIX expanded, the need for Cintas uniforms and more would grow.

Matthew granted Adam Schmitt an interview mainly because he knew him and his family on a personal level. "I wanted to give him an interview out of respect," Matthew said. "And to officially rule him out since he had no experience in our industry or anything relevant to it."

Schmitt, a big, bear of a man with a pleasant smile and a baritone voice, grew up two blocks from the welding shop. "I remember as a little boy coming up to the shop," Schmitt said. "If I broke something, I'd get it fixed before Dad got home."[78] Matthew graduated from North

Posey High School a year ahead of Schmitt. During high school and college at the University of Southern Indiana, earning degrees in human resources and business management, Schmitt worked at Poseyville's J. L. Hirsch Company Grocery and Department Store, a cornerstone business in the community, owned by the Hirsch family.

His eight years working at the Main Street corner shop, which offered everything from their notable sausage and bologna to women's blouses, exposed him to what it is like to run a family business, currently owned by Susan Hirsch Weatherholt. He learned much from Susan and her father. Hirsch's celebrated 100 years in 2015 and was once known for the quaint ritual of female butchers weighing babies on the meat scales and posting their photos on the grocery wall, a tradition still celebrated today.[79]

After college, Schmitt worked in the medical administration field. He had an opportunity to advance with that company, but it would have required him to move, and that did not appeal to him. "I'm partial to the Southern Indiana area," Schmitt said. "So, I started looking for other opportunities, and my brother forwarded me the Facebook post advertising the NIX sales position, and I thought I would try sales."[80]

After the first round of interviews, Matthew, Lindsey, and Adam were impressed with Schmitt and put him through to the second round. "After that round, we found him to be the most impressive candidate," Matthew recalled. "We took a leap of faith and hired him. Looking back, it was Adam Schmitt who was taking the larger leap of faith. He was leaving a solid job with a national company to work for a tiny family business in Poseyville. We had less than ten employees at the time, no corporate structure, and little track record for growth, other than a metal yacht and the e-house project."

"I thought, *He's from Poseyville,*" Lindsey recalled. "*He's not going to let us down. He's one of us.* I just had a good feeling about it."

Nix Metals was featured in *Evansville Business* magazine in 2013.

"I went from a company with thirty thousand employees across the nation to a company where I was one of ten," Schmitt said. "It was a struggle. I remember not knowing my full compensation package when I accepted the offer. I knew my salary, but as far as commission, I had no idea. But I knew I wanted to be back in the family atmosphere because I loved it. Corporate America wasn't for me. Too much bureaucracy. Knowing Matthew helped. I saw what he had done and what he was able to do."[81]

Schmitt admitted that he may have "pulled some wool over their eyes" when he interviewed. "One of my degrees is in human resources," Schmitt explained. "I know the interview tactics because I did them myself. I sat in the room for the first interview and thought, *I don't know how this is gonna work out with me not knowing how to weld*. I still don't know how to weld. They don't let me even touch the machine."[82]

Welding was not a prerequisite. Schmitt was hired for his business savvy and his down-to-earth, easygoing style to sell the new powder coatings and painting business as well as fabrication (developments that will be covered in the next chapter). The coatings facility was under construction when Schmitt was hired, and the first e-house project was wrapping up with a second on the horizon.

Schmitt later asked Matthew and Lindsey about their interview tactics. "It was hilarious," Lindsey recalled. "After we hired Adam Schmitt, and we laugh about it now, he said, 'Where the hell did you get that interview? I knew you two didn't come up with those questions.'"[83]

Schmitt's office at Nix Metals was quite different from the one he left in corporate America. "I went from a very large office with a couch and nice wooden desk and bookshelf to a desk from the sixties or seventies. Walking in, it was like a warehouse feel. It smelled of dirt and oil from all the gears and bearings."

A huge sliding barn door remained from the building's former occupant, Tri-County Equipment (a story covered in chapter 21). "During the winter, the wind whipped through, so I wore long underwear every day," Schmitt said. "I got creative and folded up cardboard to shove in cracks around the door to try to block the wind. But, you know, I've been blessed because I've been able to see from the very humble beginnings to where we are now, and we're still just getting started."[84]

From fake wood paneling to drywall, Schmitt remembers each renovation to move to the next level of progress and draw top-notch people to work at NIX. Matthew jokingly said Schmitt's role became doing everything that Matthew didn't know how to do or didn't want to do.

Today, as Nix Companies' vice president of business development and administration, Schmitt good-naturedly admits that this is still the case; he leads the sales and marketing team, new product and service development, human resources, and IT and data systems, along with performing many other administrative functions.

In the early days after Schmitt joined the ranks, the team tried to manufacture all sorts of finished products to keep the shop busy. Some sold well; some did not. "We never really hit any home runs," Matthew said. "Over time, we settled back into organically growing our core commercial and industrial fabrication and coatings business."

They fabricated and sold aboveground storm shelters, the closest Nix Metals came to creating a product. They tried to market tillage tool hitches (used to pull a second tool behind a disk or vertical tillage implement) with a regional direct-mail campaign, follow-up phone calls, and Schmitt traveling to every ag dealer in a several-hundred-mile radius to make a pitch. They also purchased a booth at the National Farm Machinery show in Louisville to promote the hitches. Several units were sold, but the endeavor never took off. The greater value was the experience gained.

Next, a casual comment leads to evolution.

## CHAPTER 17

# The Paint Shop

In December 2012, Matthew and Lindsey escaped Indiana's blustery winter for Florida and spent time on *Passage* with J. Z. Morris, now a comfortable friend and mentor. Matthew casually told JZ about his new idea—a paint and powder coating business.

The Nix Metals commercial and industrial fabrication business had increased and often required the subcontracting of powder coating and liquid painting, so Matthew decided NIX should open its own paint and powder coating shop.

"In those days, we would fabricate all day, then cover the machines with plastic and paint until midnight," Matthew explained. "In the winter we cranked up the heat in hopes that the next morning the paint would be dry enough to move outside so that we could get back to fabricating again. This method restricted our growth. Plus, when we outsourced the painting, the shops we hired were unreliable with regards to quality and timelines."

Matthew was merely soliciting feedback from JZ. He received much more.

"If you ever need any financial support, I would be happy to help," JZ said as he steered *Passage* past swaying palm trees.

Matthew was taken aback. The thought of an investor had not crossed his mind; his father, who was still part owner, had never borrowed money for the business and would not be a fan of taking such a risk. Yet when Matthew returned to Indiana, he entertained the idea of a partnership with JZ and soon after approached his friend to discuss. Within days, they struck a deal as business partners.

"How much money do you need?" JZ asked.

"I'm not exactly sure, but I think $250,000 will cover it."

To Matthew's surprise, JZ offered to put up 100 percent of the money as an interest-free loan and, for his capital, take a 49 percent stake in the company. This was a business lesson for Matthew. Years later, when the two men worked through JZ's buyout, Matthew asked, "Why didn't you push for fifty-one percent of the business?"

"Because if I had wanted to run the business, I would have done it on my own. I wanted you to wake up every day and go to bed every night knowing it was *your* business and the success or failure was on you. Owning forty-nine percent of that business is worth a lot more to me than fifty-one percent of a business I know nothing about and have to manage."

The partnership was even more unusual in that for a few years, there was no contract stating how JZ would be repaid. "When our attorney drafted the documents, JZ legally waived his rights to opposing counsel, saying he didn't need any," Matthew recalled. "He did not request an employment contract, which meant if I wanted to, I could have legally raised my salary or bonuses, giving all the profits to myself, leaving him nothing."

A couple of years after the coating business was founded, JZ was cruising aboard *Passage* when he shared a conference call with Matthew and his attorney and explained his thinking. "If I didn't trust Matthew, I wouldn't have gone into business with him. A piece of paper won't change that."

"It was a brilliant move," Matthew said, looking back. "I never had the courage to raise my salary one time in our partnership. And although we grew the company significantly, not until the very end did I take a modest bonus—with JZ's permission, of course. If we had drafted legal agreements stating how and when I got a raise, I surely would have taken what we had agreed to, and it likely would have been more."

In the end, the project required about $500,000 rather than the $250,000 Matthew had originally estimated. It was his first lesson on budgeting capital projects: "Take what you initially think and double it," Matthew said. "Each time I went back to JZ for additional money, he came through."

Another lesson was on resourcefulness and ingenuity. Matthew's original plan was to utilize an existing post-frame building for the coatings facility. He wanted to renovate the building rather than construct from the ground up. As the facility plans came together, it became apparent that the building, at forty-eighty by fifty-two feet, would not be large enough. However, it sat immediately next to the fabrication building, where Matthew felt the coatings facility should be located. Rather than spend money to demolish the building, he elected to move it. Once again, his father and grandfather said, "He's just young and dumb enough to do it."

Over the course of a few days, Matthew and a small team of employees cut the building's posts off at the ground with a chain saw, poured new concrete piers at the new location, temporarily braced the building to withstand the move, and dragged the building roughly 200 feet with a skid steer (track machine).

"Dad thought I was crazy for moving the building," Matthew said. "About a week after the structure was in its new location, a huge wind storm came through in the middle of the night. One that rattles your windows. I got out of bed and drove down to the shop in the pouring rain. I expected to see the building strung all over the field behind it and have to listen to my dad say, 'I told you so.' But to my surprise, it was still standing strong. Since then, we have finished out that building, and it is the cheapest square footage we ever acquired."

Matthew started construction of Nix Coatings in July 2012, and it opened for business in November. Right out of the gate, the new venture took on the painting of an aluminum fire truck tank that Nix Welding had previously fabricated and installed. It was a fiasco.

"Due to inexperience, we used the wrong materials, and the paint job turned out terrible," Matthew said. "We sanded and repainted the entire bed three times before it was worthy to send to the customer, and even then, it was not up to the standards of what we produce today. This was a very costly and embarrassing mistake. I learned that our hunger and drive would take us just so far; we needed expertise to carry us through."

Expertise was found through supplier relationships and the people the team could call for advice. Nix Companies continues to deploy these strategies today. The team was and is scrappy because early on they had to learn to be resourceful.

The next challenge: Matthew's bandwidth was shrinking. He was spread too thin. Revenue was steadily climbing at both the coatings and the metals business, and consequently, so was head count. While this was good news, it was difficult for Matthew to manage it all. Nix Coatings was not yet profitable, and after a little more than a year of struggling, Matthew called JZ and said, "I think we need to hire a manager to help oversee the coatings operations."

By this point, JZ had invested close to half a million in the start-up and had not yet seen one penny of profit. Matthew expected JZ to scold him, tell him to work harder, figure it out. Instead, he calmly listened to Matthew's reasoning.

"I have a person in mind who I think would do a great job as manager," Matthew continued.

"Okay. If that is what you think you should do, let's do it."

With JZ's blessing, Matthew got busy trying to hire his first operations manager: Brian Merkley, who was not even aware that he was on Matthew's radar. Matthew had been thinking about hiring Brian since November 2012, before his paint shop partnership with JZ had been discussed.

"The idea to hire Brian came to me as my brother, Adam, and I were walking down Las Vegas Boulevard," Matthew explained. They were there for the Fabtech Expo, North America's largest metal forming, fabricating, and welding trade show, shopping for their first CNC plasma table.

"Not that we could afford anything at the show—we already knew we would be buying a used piece of equipment—but we wanted to learn more about this technology and what manufacturers we liked before making a purchase. Besides, what guys in their early twenties would pass up an opportunity to go to Vegas for a 'work trip'? This was the first real trip of this kind my brother and I took together."

For some reason, as the neon lights of Vegas shone upon them, the two began talking about their friend Brian Merkley, who had been in Adam's class at North Posey High School and attended the same church as the Nix family. Brian and Adam had played sports together. Brian had a degree in construction management from Indiana State University and was living in Illinois, working as an operations manager for Archer Daniels Midland (ADM) Grain.

According to ADM's website, the company is "a global leader in human and animal nutrition and the world's premier agricultural origination and processing company."[85] The company has about 50,000 employees. Brian worked with grain elevators and had participated in their Grain Terminal Operations Management Program, an initiative to train superintendents to operate an elevator with the idea that one needs to know how to run an elevator to understand how to build one.

Whenever Brian returned home to Poseyville to visit, he would connect with Adam and Matthew. He stopped by Nix Welding and saw how the business was expanding.

The brothers knew Brian wanted to move back to the Poseyville area. As Matthew and Adam soaked in the dry desert heat on that November day in 2012, they tried to think of businesses around Poseyville that would be a good fit for Brian, to help their friend return home.

Suddenly, an idea clicked with Matthew.

"Why the hell are we helping Brian go work for someone else? He should be working with us!"

Adam was a realist. "He won't come to work for us. We can't afford him. Besides, we don't even have a place for him."

Matthew knew Adam was correct in that Nix Metals could not afford Brian and did not have a place for him. At the moment.

"From that day forward, I vowed to fix both of those issues, and I never believed Adam's first statement," Matthew said. "Brian *would* come to work for us. I am all about win-wins. If I truly believe something is in the best interest of both parties, I *will not* take no for an answer."

Ten months later, *Passage* would again work her magic.

## CHAPTER 18

# An Opportunity to Build Something

Matthew knew of Brian Merkley's intelligence, work ethic, and character. "In my mind, if he could manage a grain elevator, he could manage a paint and powder coat shop," Matthew said. He began wooing Brian. First, he visited him in Northern Illinois for a weekend in February 2013. Then, in April, he flew him to Florida to spend a weekend on *Passage*.

As they cruised from where JZ had *Passage* docked in Sarasota, Matthew broached the subject of Brian coming to work for Nix Coatings. At that point, JZ still owned half of the coatings business. Brian, a no-nonsense man with eager, friendly eyes, recalled, "We knew that Joe Morris wasn't going to be involved years down the road. At some point he would want out."[86]

Sitting on the yacht, drink in hand, Brian looked at Matthew and said, "I'll come on board under one condition. If JZ ever sells his shares, I would have the first right to purchase them."[87]

Matthew readily agreed, and the two men sealed the deal with a handshake. Brian started working for Nix Companies as general manager of Nix Coatings in September 2013. Several years later, he became a shareholder of Nix Companies Inc. (NCI), the parent company that now owns the coatings business. As NCI's first shareholder, Brian set the precedent for all members of the executive team to be shareholders in the company.

Like Adam Schmitt, Brian believed in Matthew and his mission to build NIX, so he accepted the risk of leaving the security of a large, established company to join. "I took a pay cut," Brian said. "I wanted to move back home. And at that point at ADM, I worked with a bunch of people I didn't really know. At NIX, I got an opportunity to work with my buddies and really build something."[88]

Working with friends is not all smooth sailing. That is a good thing. Heated discourse, questioning, and differing points of view establish astonishing power for an organization. The freedom to give an opinion and learn from others builds an open environment for ingenuity and innovation. When people know they can diplomatically voice their thoughts, they like their jobs and their colleagues. While a decision may not always go their way, they feel heard and valued. That's what Nix Companies is about.

"It's been good," Brian said. "Of course, we have our moments. There's always a little back-and-forth. But we're guys; we'll forget about it. We can yell at each other and be friends."[89]

Lindsey, who participated in the interview and hiring process of both Adam Schmitt and Brian Merkley, looks at a candidate's family dynamic. "I credit the wives, too, because if my husband had a really good paying job, and he told me he was going to go work for a start-up, have less vacation, work more hours, and probably make less money, I'd be like, uh, *no*."[90]

"And they all have wonderful spouses," Lindsey continued. "They all have been on board and have seen the vision. Matthew does a good job of including them."[91]

Today, Matthew sends a congratulatory card to each married staff member and their spouse on their wedding anniversary, thanking them for their example of commitment to marriage and family. Children are brought into the fold in many ways, including the Kids' Dollar Bill program, implemented by Angela Kirlin, director of human resources. When young children of NIX employees have a birthday, they receive a card with a dollar bill inside. The yearly company picnic brings NIX families together in the out-of-doors. The Christmas party is where the company pulls out the stops to celebrate the hard work of its people. A dinner is followed by the bestowing of sought-after prizes, including the gift of time—a giveaway of PTO, personal time off—plus a dance party with a band or live DJ.

Within a few months after Brian was hired, the losses accrued by Nix Coatings had been stabilized, and by the end of the second year, the new division was solidly in the black. Under Brian's leadership, the revenues of the coating business nearly doubled, made healthy profits, and it was possible to begin to repay JZ's loan.

After Nix Coatings was up and running for a few years, potential customers began requesting sandblasting and painting out in the field. "Not being a group to turn down opportunities, we quickly sprang into action," Matthew said. "With a truck, trailer, and essential equipment, we were in the field coatings business. The only trouble was, we didn't have a dedicated field coatings staff, so we borrowed people from the shop to go out in the field. We were constantly 'robbing Peter to pay Paul.' Eventually, we decided to go all in on field coatings."

This time the team did not wait to find a manager for the new division. "We caught wind of a young go-getter, Tyler Greathouse." Matthew said. "He was running a local spray-foam-insulation division of a regional company but was thinking of starting his own business. His skill set was transferable to what we needed. We approached him, and it was obvious he was a hardworking entrepreneur; however, we could provide him with the support and resources that with his own start-up he would have to build from scratch. After a few months of negotiating, Tyler agreed to partner with us instead of starting his own business."

When Tyler was hired, Brian Merkley moved into his current position as vice president of operations, overseeing both Shop Coatings and Field Coatings. He's also responsible for the fleets—all the trucks and their equipment, maintenance, purchasing, and selling—while leading the maintenance program for the entire company. Brian's excellent boots-on-the-ground leadership style and problem-solving skills earned him high regard with his team members and a track record for results.

Today, the Coatings Group is a stable and consistent contributor to the bottom line. The control of fabrication projects from design to finish paint provides the company with a strategic advantage that many competitors cannot offer.

In addition to his day-to-day job description, Brian took over the additional task in 2020 of overseeing a large facility expansion—a new addition to the Custom Fabrication division, housed in a newly designed building on Frontage Road in Poseyville. It's a striking building, with its roots in the wooden blacksmith shop Charles Nix opened in the center of town at the turn of the twentieth century. At the time of this writing, the company is building a brand-new home for the coatings operations, the largest facility expansion to date, four times the size of the previous shop.

Matthew and the team value Brian's "hard-nosed, no-BS approach"—although it does get him into trouble from time to time. Brian said he works on his delivery to soften his frank comments.

"I don't like to get sold on stuff," Brian explained. "Just tell me straightforward. Whether it's good or bad, I want to know. Because I will just straight tell you what I am thinking. I don't sugarcoat it. I'm not one to put anything on the back burner. I get things done, and Matthew doesn't have to worry about it. I'm doing what I do so he can think more big picture."

"Brian challenges me to be a better leader," Matthew admitted. "He's the guy I know I can go to when I want the unfiltered truth, no matter how ugly it might be. Brian reminds me of my dad the most of anyone on the executive team. Brian only talks when he thinks it will add value. He's not philosophical, and he does not like planning too far into the future. Both characteristics contrast with me. In fact, those kinds of conversations make him squirm. He's down to earth and black and white. At first this frustrated me, but I, along with the entire team, have come to value his perspective. There's no doubt our company is *much* stronger and healthier today because Brian is a big part of it."

Brian and those colleagues who also came from corporate jobs sifted through what they had learned at their previous positions, handpicked what they liked, and implemented the cream-of-the-corporate-world methods at Nix Companies. "Rather than the structure already being set when I arrived, we developed it along the way," Brian said. "We talked about what we liked or didn't like and built processes off of that."[92]

As Nix Companies hummed along with go-getters on board and progress intensifying, Matthew's aunt Caroline, who still handled administrative duties, dug in her heels, wanting everything to remain status quo—for the company to remain like the welding shop of yore. And with that, Matthew would soon experience the worst day of his career.

## CHAPTER 19

# With Me or Against Me?

As Matthew's role in the business grew along with his ambitions for expansion and change, his relationship with Aunt Caroline, the company's administrative arm and his godmother, began to deteriorate.

The two had always been close. Part of Caroline's unwritten job description had been to shuttle Matthew and his brother to after-school sports practices and other extracurricular events when they were young. Caroline left her tiny office in the cement-block building that housed the welding shop, picked up the Nix brothers from school, and carted them to wherever they needed to be because Bill and Donna were at work. A lot of quality time can be shared inside a car.

Over a three-year period, as Nix Coatings was built and Lindsey joined the company, the relationship between Matthew and Caroline eroded. "I don't recall many specific issues," Matthew said. "Lindsey's arrival only compounded the tension, and although it truly never was the intent, I think Caroline felt she was being forced out. I was simply preparing to grow, and I assumed everyone could see what I saw. That we were going to need both Aunt Caroline and Lindsey in a big way, in very short order."

Matthew looks back and sees that he and his aunt were operating on two different paradigms—scarcity vs. abundance mindsets. With the scarcity mindset, a person believes that all resources are finite and have to be divided. Those with an abundance mindset believe that resources are infinite and the pie can always be made bigger.

Caroline is a leader. She likes to be in control, whether at work or cooking in the kitchen. Her knack for taking the lead was at times comforting to those following, but when she felt vulnerable, her control often kicked into overdrive and could turn a situation sour. At age twenty-eight, Matthew was making the NIX pie bigger.

"Unfortunately, the more I pushed the envelope, the more tense things got," Matthew said. "Looking back, I can more easily put myself in her shoes. While I cannot ever totally understand what she was going through, I can empathize with the fact that all the change must have been hard for her."

Sonny tried to tell his daughter to accept the changes. "This is the way it is going to be," he told her. "Matthew is stepping on board, and we're all going to have to work as a team."[93] But even Sonny's words of wisdom did not change Caroline's thinking.

"I was young and ambitious and by that time had fifty percent ownership of the business, so I was hard-charging ahead," Matthew admitted.

Word got back to Matthew from both a major supplier to the business and a former employee that Caroline was speaking negatively about the changes. The pushback from Caroline became too much, and on a Friday afternoon in April 2013, Matthew knew, as tough as it was, he had to speak up. Matthew and Caroline stood in the opening of the large overhead door that faced away from the shop, as if neither wanted to acknowledge the shop behind them as an anchor of their past.

"Caroline, I am moving this business forward," he began.

She froze, then cocked her head, listening.

"I just need to know, are you with me, or are you against me? There can't be any in-between."

Caroline looked back at her nephew. Silence parked itself between them. After a long, uncomfortable pause, Matthew said, "I guess I know your answer." She walked away—leaving behind the welding shop, Nix Welding, Nix Companies, the family legacy—and went home without saying another word to her godson.

Matthew stood there in the cavernous silence, devastated. "I went home that evening so crushed I couldn't leave my house all weekend. I was physically ill and at times in tears. I knew how cataclysmic Caroline's leaving would be for the family, yet I knew it had to be done. I came to the realization that I had been prolonging the inevitable because of my desire to save the family dynamic. Once I understood we would continue to fight for years to come, and we had a better chance of making amends outside of the business than we did working together, the decision was clear."

Matthew believed that Caroline would have preferred to do what she had "always done," but the old ways were already gone, whether she stayed or left. The business had evolved, and she refused to evolve with it. "It was no more her fault than it was mine," Matthew said. "We were simply two people on two different paths, and there was no longer a way forward together."

Before that day in April, Matthew had talked to his father about letting Caroline go. Bill understood his sister's way of thinking; he had always relied on her to keep the books and the company afloat without owing anyone. "[You've] got to know Caroline," Bill said in 2023. "She's pretty tough and hardheaded. She was taking care of the books, and she knew how much money went for this and how much money went for that, how much money is coming in. We never borrowed a dime whenever she was there. We always paid for everything we had, and that's not the way now. The freaking payroll the company has now blows my mind."[94]

Bill saw that Caroline was trying to keep everything as it always had been. "They're fighting like cats and dogs, and you can't have that crap, and somebody's got to go," he explained. "Caroline wasn't going to change her attitude or the way she thought about things. Letting her go really pissed off part of the family. But it's better now."

Caroline saw it differently. She said it was more like "butting heads." She felt like Matthew was keeping his actions regarding the company from her and went behind her back to accomplish whatever he wanted. Unbeknownst to Matthew, she even talked to Bill about it.

With gratitude, Matthew acknowledges that his father backed him up. "That must have been really hard for Dad. I hope I am never put in that situation."

With a sense of foreboding, Matthew talked about it early on with his mother. He predicted, "It's not going to be good, Mom."

"He didn't want to do it," Donna said later. "But he had to. Bill was on board with it. And that's his sister. So what do you do? You just brace yourself and hope for the best."

Matthew's brother, Adam, looks back on those days with great sadness and wonders if he should have done more, but he was new to the company then and felt like it was not his place. He understood both ways of thinking about the business—the conservative mindset and the growth mindset. He "hung out in the middle."

"There was always a bit of a tension in the air. I knew enough at that point where I realized there was something brewing, and I didn't like it. It made me feel uncomfortable. I just wanted to come in and do my job.

"Aunt Caroline helped me out tremendously growing up through the shop," Adam continued. "She went to bat for me a lot of times when Dad called me out on the quality of my work. Things got heated, and there would be arguments. Caroline always did a good job playing referee and erring on my side. We had a close connection."

While Lindsey's presence seemed to exacerbate Caroline's discontent with the changes in the company, the two had been quite close prior to Lindsey joining in May 2012. "When Caroline was ill while I was in college, she was going through some treatments in Indianapolis," Lindsey recalled. "I drove from Bloomington to see her and spend time with her. When Matthew and I bought our first house, she and her husband, John, helped us paint. When I came to work for the company, I thought, *Oh, this is great. We have this great relationship. We're going to have so much fun working together.*"

Lindsey was originally hired to sell for the then newly established

Nix Coatings and said that Caroline was "very mothering" when it came to what Lindsey and Matthew were doing regarding the changes they were making. Caroline's "mothering" came from a place of caring, yet she hovered. "I think she thought my coming in with these new ideas was threatening to what she was doing," Lindsey said. "I was just trying to help. I wanted nothing to do with accounting. I wanted to be out selling the coatings business so the company could grow."

For decades, Caroline had worked out of the closet-size office in the cement-block building Sonny and his father, Carl Sr., had built. Matthew and Lindsey saw that she never had space to work and moved her into a nicer office across the street in the building on Fletchall, where the coatings and burgeoning fabrication divisions were located. "We thought we were helping her with a better environment," Lindsey said. "But I think she liked being back in the old shop with customers. It was too much change, and she wasn't on board for that."

The day Caroline left, Matthew looked at Lindsey and said, "You're going to have to do her job."

Lindsey was completely agreeable. She said, "I can do this." Later, she explained, "I knew to do what we wanted to do, Caroline was going to keep holding Matthew back."

Lindsey immediately called their accountant and said she needed someone to help her run payroll that week. "I started doing all the billing, paying all the bills, doing all the payroll, everything. Plus still trying to market the coatings business." She ended up doing the job for five years and giving birth to two children during that time.

Lindsey has had type 1 diabetes since she was sixteen. Her pregnancies were considered high risk. "When we had our first son, Charlie, who was born in 2014, Matthew brought me time cards and statements at the hospital. I had Charlie four weeks early. I was folding statements and stuffing them in envelopes. It had to be done because if the money is not coming in, we can't pay our employees. The nurse walked in and said, 'Can't someone else help you with this?' I said, 'No, I'm it. It's just me.'"

"Payroll has to happen," Matthew explained. "It doesn't matter if you're in the hospital with your newborn. That's the type of little nuances of owning your own business, particularly being in business with your spouse, that people may not think about."[95]

While handling the accounting from a hospital bed, Lindsey and Matthew worried about their premature firstborn. Charlie was ventilated immediately after he came into the world. Though it was scary at first, he turned a corner and was released from the hospital in about seven days. There was no maternity leave for Lindsey; she took little Charlie to work with her and worked some from home.

A business often has growing pains—spats among its people, hurt feelings, miscommunications, differing opinions on how the ship should be steered. Some employees leave or are pushed out. A family business takes those intense moments to a higher realm of emotion. How do you love someone who fires you or questions your work ethic or the quality of your output? How do you sit down in front of a turkey and eat cranberry relish with the person who yelled at you about the bottom line? Matthew did not fire his aunt. He did not fire his godmother. He fired his bookkeeper.

"What I hate is, gosh darn, why did that have to be part of our history?" Adam said. "Because you hate for it to be documented. You just hate it."

"Someday, I too run the risk of not evolving," Matthew admitted. "And if I do not, I expect that the business will need to leave me behind as well. There will come a day when I am no longer the most qualified to be the captain of the ship. I hope I recognize that before everyone else does, but only time will tell."

Letting Caroline go was the hardest decision Matthew has ever made. "I genuinely always wanted her to come along with us on the growth journey. She could have been incredibly valuable. There were many times over the years that I thought to myself, *I wish Caroline were here to help with this. She would be a great asset for this.* But given the foregone circumstances, I know I made the right decision, and I hope that she is happier now too."

It took about ten years after Caroline left for the family to come together and be cordial at holiday dinners. "It's better now, but it has taken a long time," Lindsey said. Caroline enjoys being around Matthew and Lindsey's three sons, just as she does all her great-nieces and nephews. "Unfortunately, sometimes it takes hard times to bring families back together," Matthew said. "We lost an uncle who was Caroline's brother-in-law and Grandpa Sonny. I think those things helped us to put some of that stuff behind us. We are still family. I still love her. That will never change."

# CHAPTER 20

# Learning to Lead

When Matthew became the shop foreman, he learned to lead others by trial and error, which often resulted in hard-wrought lessons in humility. He was a blue-collar part owner who had never led employees or experienced formal leadership training.

"My only form of 'training' was the authoritative and hard-nosed directives from my dad and grandpa. Yelling and profanities were the norm. When I questioned means, methods, or reasoning, the common answer was, '*Dammit*, because I said so!'"

Matthew's early style mimicked that of Bill and Sonny because that was what he knew. And he paid the price. Attendance, quality, honesty, and dependability were problematic. "However, had I not been tough, I would have gotten walked on," Matthew recalled. "It was a crude environment in which to learn to lead."

The atmosphere in the shop was rough-and-tumble. A stark contrast to the pleasant, friendly atmosphere at Nix Industrial today, where the caliber of employees is top notch and Matthew and the executive team strive to lead with consideration, grace, and diplomacy.

On one occasion, an employee stole a new torch hose, and Matthew drove to his house to confront him. Fortunately, the man was not home. But as Matthew left, he saw the hose in the back seat of the employee's car, opened the door, and retrieved it. The employee never showed up again for work, and Matthew did not file a police report, believing karma would take care of the culprit.

"In the early days, I didn't do a good job of hiring or holding people accountable to a consistent set of standards," Matthew said. "I didn't understand what motivated people. Eventually, I learned to slow down on the hiring process. I also learned that letting people go is often the best answer. I call it 'pruning the branches.' Today, I remind our team that letting people hang around who don't align with our values and don't fit our ethos sabotages the great culture we've built."

Today, Nix Industrial's fundamental value is "caring for others." Matthew explained:

> Bending the rules always comes back to haunt the company. I'm not talking about failing to extend grace when it is deserved. That's something totally different. I'm talking about sweeping bad behavior under the rug because we don't want to deal with it. If a good employee is going through a hard time, we should always treat them like a human, not a unit of production. If we don't give people space and time to deal with life's challenges, we are not living up to our core value—caring for others. We also won't have many team members left.
>
> But if someone habitually has issues, that's a different story. Eventually people have to take responsibility and help themselves. We do them and the business a disservice by allowing them to hang around. It never gets easier, but I've learned the hard way that we usually regret delaying the necessary action to let an employee go.

Matthew also learned that more could be accomplished and he would gain respect when he explained the "why" behind an employee's work.

"In the early days, when we needed to make schedule or priority changes, I would go into the shop and say, 'I need you to stop working on that project and start working on this one.' They would typically be frustrated by the change, and that in turn frustrated me. I thought, *Why do they care? They get paid the same either way. Why don't they just do what I tell them and let me worry about the schedule?* I've since learned that most people do care. It's helpful for them to be able to connect their work to the 'why.'"

NIX wants the kind of people who want to know why. They don't want yes-people who follow like sheep. They want employees who think for themselves and try to find ways to improve procedures and practices. When Matthew explains that a project is the new priority because, for example, a good customer asked if it could be turned around quickly so their manufacturing line doesn't shut down, employees happily switch gears and make it happen. They connect with the "why" and find fulfillment in the shift of priorities rather than frustration in not finishing what they started.

Matthew also learned that one way to motivate and garner an employee's buy-in is to ask for input. "I try to make fewer statements and ask more questions. This is something I'm still growing into," Matthew admitted.

During the early explosive growth of the company, when the second contract of e-houses was in full swing and the construction of the new coatings shop was underway, Matthew was spread thin—working on the shop floor by day, handling estimates and invoices at night, and starting a new business. He needed to appoint a foreman, fast. He walked over to Tony Shell, handed him a key to the shop, and said, "I want you to be the foreman."

"I took the best guy on the shop floor and made him the foreman," Matthew said. "There wasn't any training or any formal promotion like we do today."

However, Matthew did not immediately stop working in the shop. He came and went for a few years after Tony was made foreman. This proved to be a lesson in how *not* to lead. Matthew inadvertently undermined Tony's authority. "By constantly coming and going from the shop floor, I was making it harder for him to do his job, take full responsibility (because he had me as a crutch), and gain the full respect of the people he was tasked to lead."

Matthew's evolution from welder to administrator was gradual. He slowly spent more time in the office than the shop and stopped wearing welding uniforms in 2014. "I still keep a uniform shirt and welding hood in my office," Matthew said in 2023. "A few times a year, I go out and work in the shop. It's like therapy for me, and it shows the new people I still remember how to weld. Some of the young workers assume I always had an office job, and they are shocked when they see me out there."

Over time, Matthew learned to trust Tony and let him take full charge. Matthew coached him on how to lead, lessons he had learned the hard way. He came to realize that Tony was the guinea pig, enduring many of the fledgling CEO's early management mistakes, and at one point considered leaving. Matthew apologized.

"Fortunately, we talked it out and put it behind us," Matthew said. "I'm now convinced he never wanted to leave in the first place. He was frustrated and wanted my attention to make corrections."

Gradually, Tony took on more special projects and was promoted to a management position. He helped move the business forward in multiple ways and became Matthew's close confidant.

In 2018, Nix Companies had a field installation in the farthest location from Poseyville to that date—Santa Clara, California. Matthew sent Tony to run the job site and oversee the subcontracted steel erectors. It went so well that the customer requested Tony for his next project.

In March of the next year, Tony came full circle. He oversaw the large-scale renovation of the new Frontage Road home for the Custom Fabrication division, a building that replaced the old pole barn where Tony had originally started working for Nix Companies. Where Matthew once handed him a key and said, "I want you to be foreman."

• • •

## SONNY'S PRIDE

After Sonny retired in 2009, he distanced himself from the business. Matthew had taken on a leadership role but was not yet formally in charge. "Due to the tensions caused by our growth, Grandpa and I weren't on the best footing when he retired. So, I took it personally that he didn't come around. With each new growth milestone or new cool project, I wished he would come by to see it. Every young man (whether he will admit it or not) wants his father's validation, and when you grow up working alongside your grandpa, you also seek his validation."

Matthew's frustration that Sonny did not step foot in the shop after retirement was compounded by the fact that the man lived a mere block away and his garden, where he spent much of his time, adjoined the shop property. "I regularly caught him driving past the shop in his little red Chevy S10," Matthew said. "Obviously checking on us to see what was going on. When I talked to him, he acted disinterested. At least, that's the way it felt, or what I made myself believe."

Without any formality, Matthew took over as president. Bill had never referred to himself as president, even though he was, according to the corporation paperwork. Matthew recalled:

> Once Adam and I had majority control of the company [in July 2013], the corporate documents listed me as the president. For a while, I didn't adopt the title either. At some point, when I thought it was appropriate, I included it in my email signature and business cards. That was about as official as it got. A formal changing of the guard didn't happen. With each major decision that needed to be made, Dad deflected to me. While I had challenges working alongside my dad, him being a backseat driver was not one of them. Once he made the decision to sell, he allowed me to control and lead. When I asked, he gave advice or guidance but rarely intervened in big decisions. He focused more on the smaller, day-to-day details, which he never gets tired of griping about. That's just who he is, and we are better for it.

A few years after Matthew became president and the company began to grow by leaps and bounds, Matthew heard comments from people in Poseyville. "I saw your grandpa the other day. He's really proud of you." Their words were heartwarming, but Matthew was confused. Why did Sonny not stop in and see what Matthew was doing with the business?

One day, Matthew was talking to retired community legend Coach Joe Gengelbach, who led the North Posey football program for forty-two years. He had coached Bill, Matthew, and Adam, and the high school football field is named after him. Gengelbach is a member of the Indiana Football Hall of Fame.

"It was August, and football practices were about to start," Matthew recalled. "I asked him if he was going to go visit the kids at practice or help out. He said they had asked him to help every year since his retirement, but he felt like it wasn't right."

"It's theirs to run, and I don't want to get in the way," Gengelbach explained. Then he said something that changed Matthew's view about his grandfather. "You know, I was thinking that those juniors and seniors now don't even know who I am. Maybe I can finally go back and help out. No one will look at me as 'the guy.'"

Matthew immediately thought: *That must be how Grandpa feels.*

Shortly after, Matthew saw Sonny pulling weeds in his garden. He stopped and told him the Gengelbach story. He also mentioned he was disappointed that his grandfather had not come around. Sonny admitted that he was taking a similar hands-off approach to the shop. He was staying out of the way. Then he added, "I'm proud of what you and Adam are doing."

"That was the proudest day of my career at that time," Matthew said. "A giant weight was lifted. I invited Grandpa to come around more, and after that day, he visited from time to time. When we acquired our shops in Rockport and Princeton, I took him there. He attended our biannual company meetings, where he was a big star. The company had grown nearly one hundred times over, with team members spread across four states, and this small-framed, eighty-six-year-old, humble man stood up and waved to everyone, proving that we are still, at our core, a family business."

• • •

## LEADING WITH LEGACY IN MIND

At age twenty-nine, Matthew sat in the hospital room where his first son—named after his great-great-great-grandfather and company founder Charles Henry Nix and his grandfather Charles William "Bill" Nix—lay in an incubator, born prematurely. To take a break from the strain of watching his fragile newborn, Matthew made a trip to the Barnes and Noble bookstore and walked the nonfiction aisles. A book jumped out: *Leading with Your Legacy in Mind* by Andrew Thorne.

"I picked up Thorne's book, the first book I read since high school and the first book I read willingly in my adult life," Matthew recalled. "Charlie was the first potential successor and sixth generation of Nix Companies. The timing was right, and the soil was fertile. That moment began my journey of learning and thirst for information that has not stopped."

By this point, Matthew was leading professional staff members with college degrees. "In the beginning of this phase, as someone who had not gone to college and was building the company, I viewed my lack of a degree as a badge of honor," Matthew said. "I had a bit of a chip on my shoulder for the higher-education process. However, as time went on, my feelings began to shift. As I interacted with college-educated folks, doing my best to lead them, I felt lacking at times. With aspirations to continue to cultivate the business, I knew there were a lot of administrative and leadership fundamentals I could stand to learn. In hindsight, my time at Vincennes University would have been better spent in business classes."

Four generations of the Nix family, L–R: Adam Nix, Charles William "Bill" Nix, Matthew Nix holding Charles Matthew "Charlie" Nix, Carl A. "Sonny" Nix Jr.

Matthew began to educate himself through books, lectures, and college courses he found online. The following year, on his thirtieth birthday, Matthew wrote his "Before 40 Bucket List," envisioning what he wanted to accomplish in the next decade. One of his goals was to read 200 books. Buying hardback books and building a home library became a hobby. Also on that list was his aspiration to return to school and advance his formal education.

After some research, he landed on Notre Dame's six-month executive business administration certificate program, which taught much of the core materials a student learns earning Notre Dame's executive MBA. The course was short and thus intense, designed for professionals to complete while working.

"I'm now convinced it wasn't designed for professionals who are working and have two kids at home in diapers," Matthew said with a laugh. "When Lindsey and I reminisce about that period in our lives, she says somewhat jokingly, 'I don't even remember that time. I think I've just blocked it out of my memory.' I must say, I could not have done it without her support."

Beginning in 2016, for six months, Matthew worked all day, came home, ate supper, helped put his two sons to bed, then studied until 11 p.m. He studied on Saturdays and on Sundays after church. His classmates in the Notre Dame program were military officers, NASA employees, and business executives. He was the only one without an undergraduate degree. He finished near the top of his class.

"I was doing it because I had a genuine thirst for knowledge," Matthew said. "I hope that thirst never goes away. There's an old saying: 'If you aren't growing, you're dying.' If a business isn't growing, it's dying. The day we stop growing, in whatever way we aspire, that's the day we start dying."

## CHAPTER 21

# Rock Bottom

There's a concept called normalcy bias where one experiences a set of events over a short period of time and begins to think that trend is normal or to be expected. In business and in life, a short-term pattern of positive results can be intoxicating, and we might get sucked into the normalcy bias of believing everything we do will work out that well. This was the case for Matthew and Nix Companies in the fall of 2014.

The company had experienced its fifth consecutive year of phenomenal progress. The successful launch of the coatings division helped quadruple the size of the business. Much of the growth was organic, made possible by low-hanging fruit. Smart decisions had been made, but to some extent the business was also riding the wave of economic recovery after the country's downturn in 2009.

With good fortune tucked under his belt, Matthew learned that Tri-County Equipment, a long-standing Poseyville business at 160 West Main Street, was closing its doors. For more than thirty years, the tractor dealership had offered parts, sales, and service to the agricultural community. The longtime owner had recently sold the business to a small out-of-town investment group that was now struggling, liquidating assets, and closing the doors. By January 1, 2015, the eight or so remaining employees would be out of a job.

The original local owner leased the land where Tri-County Equipment stood to the investment group. On the verge of ending up with a vacant piece of property, he approached Matthew about an idea to keep the business alive.

"Don't you even think about buying that place," Lindsey said the second she knew Matthew had been approached.

"Don't worry. I don't have any interest in it, but let's hear him out."

Matthew, Lindsey, and Adam listened to the owner's proposal. Matthew was intrigued. "Despite telling Lindsey initially I was not interested, I convinced her and our banker it was a good move," Matthew recalled.

However, as the family sat in a room together, Bill had to be convinced as well.

"Who's gonna run it?" Bill asked.

Matthew looked at his brother. "Adam will."

"By December 31, 2014, we had closed on our first acquisition, and we were the proud new owners of a failing ag-equipment dealership," Matthew said. Adam began to lead the new endeavor in January, and he considered the purchase of Tri-County Equipment, renamed Nix Equipment, a smart move. Today, the divisional identity is Repairs & Maintenance, alongside Midwest Trailer Sales, the only equipment sales that remain from the original acquisition. "I thought it was a good opportunity for the company and for me professionally," Adam recalled.[96]

Lacking the experience to enter the big-tractor market, the company decided to focus on being a short-run dealer. They would sell the smaller second- and third-tier equipment lines, which are usually more profitable by margin percentage and are commonly sold by the big dealers as supplemental to their primary tractor line as a volume game.

"There were plenty of successful stand-alone short-run dealers out there," Matthew said. "But truth be told, we didn't study them to the extent that we should have, and, unknowingly, we were undercapitalized. We decided to merge our legacy business [ag, transportation, and construction equipment repairs, a.k.a. the welding shop] that Adam and Dad ran with the newly acquired ag-equipment business. Since they shared the same customer base, the merger would make that division more substantial and take advantage of synergies.

Adam had been wanting more responsibility, so this placed him in a formal leadership role as the shop foreman and service manager, which included a few mechanics and welders to oversee, while still working in Dad's shadow. It was a lot to juggle."

Matthew's primary play was for the real estate. "Our core business was growing, and this was the largest piece of commercially zoned property in town or the surrounding area," Matthew explained. "Plus, the previous owner offered to sell it on contract with no bank loan needed. It seemed like a good move. We just had to find a way to cash-flow it."

After all, Nix Companies was on the climb up. The "new normal" was fast-tracked success. Why would this undertaking be any different?

After the first year, most of the Tri-County inventory had been sold, but Nix Equipment still struggled to meet the volume and margins necessary to make it profitable. There was not enough work to keep four mechanics busy. One left on his own accord, but another had to be let go. It was the first time Adam, age twenty-seven, had to fire someone. Nix Equipment was struggling to survive.

"It was the most difficult decision I had ever made at that point in my professional career," Adam said, looking back. "The hardest part was the fact that Harold didn't do anything wrong. We simply did not have enough work to keep all four mechanics busy. I remember the moment vividly. I let Matthew know I would take the lead since Harold reported to me, but I wanted Matthew there for support."

Matthew coached Adam before they met with Harold, reiterating that it was the right decision. He told Adam to stick to the facts so he would not become emotional.

Harold came into work that morning expecting a normal workday. He had even pulled his camper to work because he was heading to his campsite after his shift. Adam called him into his office at the repair shop at 7 a.m. and let him know he would no longer be employed by Nix. Matthew quietly sat in on the meeting, letting Adam lead.

"Harold took it relatively well and understood the decision we had to make," Adam said. "It didn't make it any easier. It was one of the worst feelings I have experienced. Sadly, I heard he passed away in January 2021. I have had to let several people go since then. Our people are our biggest asset, and we're thorough when it comes to hiring, so letting people go for any reason is never easy. I always lose sleep over it."

Matthew and the team learned a hard lesson about looking out for the business as a whole. "If it fails, everyone is out of a job, our family would be in a desperate situation, and there would be vendors and creditors clamoring to be paid," Matthew said.

Adam was "cleaning up old messes." Today, NIX takes pride in "never laying anyone off"; however, push came to shove with two early situations that came about because of acquisitions and issues resulting from a prior owner's actions that had to be confronted. Looking back, Matthew said, "Since the large tractor dealership had been lost prior to NIX acquiring the business, the downsizing of mechanics seemed inevitable and out of Adam's control, but that didn't make it any easier for him."

Soon, Adam realized he did not have the experience needed to run Nix Equipment. "This was my first big opportunity to prove myself, and I fell completely on my face," Adam recalled. "I knew early on that I was in over my head and let Matthew know I needed help."

Adam, Matthew, and Lindsey agreed that they needed to hire a general manager. An ad for the position was placed on social media. They were pleasantly surprised to receive an application from Brandon Wright, originally from Owensville, ten miles from Poseyville. He had gone to high school with Lindsey, who had been a cheerleader with Brandon's wife.

Brandon had an industrial technology degree from Purdue University and was working at Flanders Electric Motors, an Evansville company that engineers and rebuilds electric motors for heavy industry. In 2015 Brandon left his solid job in a large, established, family-owned company to join Nix Companies, essentially a century-old start-up. Today he is vice president of operations and an equity partner.

"There's a couple reasons why I moved to Nix Companies," Brandon explained. "I had been following Matthew through social media, and I was watching the business grow. The trajectory interested me. I wanted to be on the ground floor of that. My grandpa was a farmer, and I can remember going to the welding shop as a kid. So that was another thing that interested me—being part of a local business that provides jobs for our community."[97]

Initially, Brandon came in as general manager of Nix Equipment, overseeing financial and HR responsibilities. Adam ran the shop as the service manager, and the arrangement worked well. "At the time, Brandon was the most qualified and educated person we had ever hired," Matthew said.

Matthew gives a lot of credit to his brother for wanting to hire a general manager. "The scenario speaks volumes about Adam's self-awareness, humility, and priorities. As the second-largest shareholder in the company, he was essentially signing off on hiring his own boss." Today Adam leads the two separate divisions as vice president of operations.

Brandon's first years were challenging due to intense financial struggles along with cultural differences between the Tri-County Equipment way and the NIX way. "The employees that came from Tri-County were used to being told what to do," Brandon explained. "They were accustomed to an authoritarian atmosphere. They were managed, not led. They had been expected to stay in their lane. Our approach is always to get an employee's buy-in and give them leeway. Some struggled with that. It was funny because they would say they didn't like the old way, but then they tried to do things the old way."[98]

The acquisition of Tri-County Equipment included lawn-and-garden retail sales, an anomaly for Nix Companies. Second-year revenues were better, but the grind of the lawn-and-garden aspect was wearing. The stress of constant cash-flow challenges was extremely hard on Lindsey. "She was still managing the company accounting and finances as the director of finance, and we lived the challenges day and night, at the office and at home," Matthew recalled.

Lindsey went so far as to move her office to the Tri-County Equipment location to be closer to the bookkeeper who had come along as part of the acquisition. Lindsey did not *want* to leave the NIX main office and take on the challenges of the newly acquired business. She had been against buying it in the first place. Yet, like many other times, she did what needed to be done.

"I was so in the weeds of the challenges that I didn't recognize the sacrifices she was making," Matthew admitted. "Now having the luxury of hindsight, I don't think it's possible to overstate the ways in which she helped keep it all together during those trying times."

"We were selling lawn mowers and bush hogs," Lindsey described with a laugh. "We tried to sell tillage equipment. It was really bad. Brandon was great at dealing with problems and moving on. I give him credit because he took a lot of crap in that business. It was customer facing, so he was dealing with the public."[99]

On top of the new acquisition struggles, Nix Companies experienced a financial hit when the coal and oil market took a plunge in 2015. At the time, coal and oil made up more than half of the core business (Nix Metals); it was a great lesson in the value of diversification. Also, a large automotive customer was stretching out its payments, and Nix Equipment had a large inventory payment due. At one point, there was only $8,000 left in the checking account, and Matthew and Lindsey did not know how they were going to make the next payroll. To put it in perspective, at this time, payroll was over $20,000 a week, and the monthly accounts payable to vendors totaled around $100,000 per month.

"We only had a small line of credit, and the thought of asking for that to be increased seemed implausible and irresponsible," Matthew remembered. "We were in the middle of a perfect storm of financial ruin."

One evening, it all came crashing down in the couple's kitchen. Lindsey stood at the stove cooking dinner. Matthew looked at the face he had loved since he was fifteen. He saw tears of desperation in her eyes. Tears he interpreted as her disappointment in him. The sight of her pain broke him.

Later, Matthew was putting two-year-old Charlie to bed. The toddler could barely talk. As Matthew laid his son in his crib, Charlie peered up into his father's face, worn with worry. "What's wrong, Daddy?"

At these innocent, intuitive words, Matthew crumpled to the floor and sobbed. Desperate thoughts ricocheted. *What in the hell have I done? I had a perfectly good family business, and I went and f----ed it up.*

The Charlie moment was Matthew's rock bottom.

He prayed. And prayed. And prayed some more. The next morning, he awoke with resolve to correct the course. What seemed like insurmountable obstacles the day before now felt like insignificant nuisances to be worked through or around. He called Adam and laid out a plan. He kept praying.

Miraculously, a few days later, like manna from heaven, the big automotive customer paid their large invoice, and suddenly there was a tiny bit of financial breathing room. Next, Matthew called the investment group that had sold Tri-County Equipment and discussed renegotiating the terms of their deal. Some of the inventory had been financed by the seller as a part of the acquisition. Matthew told him Nix Companies might go broke and, as a result, he would get nothing.

"In hindsight, we had plenty of equity in fixed assets that were paid in full, and we probably paid cash for too many assets back then, which caused an even greater strain on short-term working capital," Matthew recalled. "Not that we wanted to liquidate the assets. We needed them to continue to operate. But had it come to that, all creditors would have been covered. So, to that extent, I misled him. I didn't mean to, and I feel bad about that, but I was acting out of desperation and thankful it worked. He agreed to renegotiate."

The Tri-County Equipment inventory was sold, some of it in a one-day online auction, and the Nix Companies seller debt was paid off. Over the course of a few months, close to $1 million in inventory and associated debt was unloaded, ridding Nix Companies of a distracting, low-margin business. The real estate that Matthew highly valued for potential growth remained a part of NIX. The business unit would concentrate primarily on service work and was rebranded as Repairs & Maintenance.

"At this point I had the opportunity to run this business unit as the operations manager," Adam recalled. "With the time I spent learning the business from Brandon, I was ready. Brandon is now lateral to me as a vice president of operations, although I still bounce a lot of questions off of him. I have the utmost respect for him, and I'm glad we had the opportunity to bond in those early years."

Matthew and the team—and, by all indications, a higher power—transformed what could have been financial ruin into lessons learned and a laser focus on the core business.

Lindsey looked back in 2023. "There have been so many highs and successes that outweigh the lows like Tri-County Equipment. I remember thinking then, *We've got to make this work, because I've known Brandon and his wife forever, and I don't want to let them down.* They had two kids at that point. But the experience set him up for future successes, and he became our acquisition guy."[100]

Tri-County Equipment became a blip in the company's story. Soon, Matthew would hear about a massive exchange of wealth that was about to occur in the country, read three life-changing books, embrace the "hedgehog concept," and position the company for evolution. His prayers had been answered.

It had been one year since Charlie asked, "What's wrong, Daddy?"

# PART 3

# Precision-Shaping the Future

## CHAPTER 22

# Search for Meaning

In 2013, Matthew was invited to become a member of a new grassroots group for local business owners that was later named Forum 4 Growth. Members currently meet monthly to share ideas. The first meeting was at the Red Geranium restaurant in nearby New Harmony. As the group sat in the bar, Andrew Wilson, a member who was in the auction and realty business, described the massive transfer of wealth the country was about to experience, the likes of which had not occurred before. As the baby boomer generation began to retire and pass on, their wealth would be transferred to the next generation.

Matthew listened intently and thought, *Wow! This guy is sitting in front of a monsoon of money headed his way, and he's in a prime position to capitalize.*

"To me it was the equivalent of being in one of those booths where the money swirls around, and you have to grab as much as you can before the timer goes off," Matthew said. "I've never been terribly motivated by money. My motivation comes from the thrill of building something. However, I understand economics. I see the value of money for what it is, a resource or tool. I saw that putting oneself in a position to capitalize on this event was the smart thing to do. But I was in the welding shop business and did not have a business model to take advantage of it. I thought, *I need to figure out a way to get in front of this wave.*"

About this time, Matthew read the book *Good to Great* by Jim

Collins. The book's idea of the hedgehog concept was life changing for Matthew. Collins writes,

> In his famous essay "The Hedgehog and the Fox," Isaiah Berlin divided the world into hedgehogs and foxes, based on an ancient Greek parable: "The fox knows many things, but the hedgehog knows one big thing." . . . Foxes pursue many ends at the same time, . . . are scattered and diffused, moving on many levels, never integrating their thinking into one overall concept or unifying vision. Hedgehogs, on the other hand, simplify a complex world into a single organizing idea, a basic principle or concept that unifies and guides everything.[101]

The model refers to the idea that some companies try to be good at many things rather than focusing on being great at one thing. Collins uses a Venn diagram—"Three Circles of the Hedgehog Concept"—to demonstrate how a business can uncover what will make it great. Each circle contains one of three questions: 1) What are you deeply passionate about? 2) What drives your economic engine? 3) What can you be best in the world at?

The point where these circles intersect is your hedgehog concept.

The hedgehog concept resonated with Matthew, and after months of deliberation, he answered the questions that would set Nix Companies on its path to greatness:

1. What are you deeply passionate about? "People and corporate culture."
2. What drives your economic engine? "Skilled labor—tradespeople."
3. What can you be the best in the world at? "Operating a network of small-to-medium-size job shops in our industry."

"Our hedgehog concept is to operate as many small-to-medium job shops in our industry as possible, as opposed to growing a single shop," Matthew explained. "There are profound differences in these two concepts. It became clear that our path was to acquire other shops as their owners approached retirement. I knew this not in a big 'ah-ha' fireworks-celebration kind of way but in a calm, matter-of-fact way. It was as obvious to me as saying the sky is blue or the trees are green. And our hedgehog concept is as clear and relevant today as the moment when it entered my mind."

It may seem ironic or contradictory for some to learn that Nix Companies focuses on being the best at "one thing," since today it is so diversified. Many outsiders—people of Poseyville who watch the growth of the company from an armchair—talk among themselves and ask questions of employees, speculating that Matthew and the team are taking on too much, expanding throughout the community and beyond and overextending. The tiny welding shop is tiny no more. Matthew said.

> I can certainly appreciate why outsiders might feel that way, and believe me, we are always evaluating if a product or service is "core" enough to our business to offer it to our customers. Those of us who are behind the curtain understand that while we are very diversified in what we offer to our customers, we *are not* trying to serve every customer. We are focused on certain industry sectors, and our model is to be a high-touch, high-value relationship for them and be their easy button. That doesn't mean we will do anything. It just means our niche is focused more on our customer than it is on our services. For every one acquisition we make, I look at nearly one hundred on paper and go visit as many as twenty in person. We are much more focused and diligent than what people understand from afar.

Matthew breaks down his hedgehog-concept verbiage regarding operating a network of small-to-medium job shops in his industry:

## TO OPERATE A NETWORK

This refers to our scalability. We have an audacious vision, and there is no cap on it. We see a tremendous opportunity to play a part in the consolidation of our industry. Not "consolidation" in the sense of shutting down businesses and relocating them. But like many industries where there was a "mom and pop" business in every town, the ownership of businesses in our industry will continue to be consolidated. Think about grocery store or drugstore chains, gas stations, and the list goes on. Our industry is about twenty to thirty years behind the curve on this, but it's happening.

For the most part, if you want to sell your job shop, you have three options: sell to a family member or employee, sell to private equity investors, or auction off your equipment. Many of these businesses are not large enough to attract private equity buyers, and even if they were, most owners do not want to sell to investors. They want someone who has a deep understanding of their business, takes care of their people, and honors their legacy.

If they don't have a family member in the business or an employee who is willing and capable of buying it, they are left with closing the doors and selling off the assets. There aren't many family businesses like us who have the ability to execute on this model. This makes us unique.

## SMALL-TO-MEDIUM JOB SHOPS

We are not looking for big businesses. Rather than owning four or five large businesses, we are looking to operate dozens, or hundreds, of smaller ones. We can drive larger value creation. Most small-to-medium businesses in our industry do not have the corporate structure or shared services [accounting, IT, HR, business development, safety, quality] to scale their business or to make it marketable to outside buyers. Leveraging our corporate structure, our shared services, and our knowledge of the industry, we can acquire these businesses and create instant value for them and us when they are integrated into our larger business portfolio. The value creation is realized by their team members, their customers, their vendors, and the broader community. Each time we acquire a business, the parent company is strengthened, and the new acquisition allows us to invest in the next one.

## IN OUR INDUSTRY

We intend to stay within the commercial and industrial industry. Our legacy business, Nix Industrial (currently around 80 percent of our overall portfolio), will continue to focus on our core offerings: metal fabrication, machining, and industrial coatings. Outside of this core business, we will pursue other ventures that are typically job-shop-service oriented. As an example, we recently invested in a heavy-commercial-vehicle body shop. This was not integrated into the legacy company but operates as an independent holding. While repairing a wrecked semi is not exactly the same as welding, machining, or coating an industrial part, the management structure, the economics of the business, and, most importantly, the recruitment tactics are transferable. This is our niche.

While the company's first acquisition, Tri-County Equipment, was a tough trial by fire that almost pulled the business under, it provided a glimpse of what could be. "Sometimes overcoming challenges gives us the confidence to try something similar," Matthew said.

He was ready to focus on acquisitions by staying within the company's lane. Matthew connected the dots between what he had learned about baby boomers transferring their wealth to the next generation and the many blue-collar owners of welding and fabrication shops throughout the Midwest who were retiring and wanted to sell their businesses.

"Just like my dad, they had built a great business using their brains and back but had not focused on their succession plans," Matthew explained. "These businesses often reached what I later coined 'maximum complexity,' the threshold to which a business can grow without removing the owner as the bottleneck for decision-making and the execution of management activities. My plan was to change all of that and introduce an entirely new opportunity to the market and these sellers."

Matthew also read that the size of a business's market cap potential is directly related to the size of the problem to solve. At Nix Companies, the work performed—welding, fabrication, industrial coatings, millwright, and maintenance—solved many problems for its customers. However, Matthew quickly realized that the larger problem the company solved was the exit strategy for the owners. This was not a million-dollar problem but a *multibillion*-dollar problem. One that no other privately owned company was solving at scale.

"But just because I knew it was our path didn't mean it wasn't scary or challenging," Matthew confessed. "I spent several months asking myself if we really wanted to go for it. Then the fourth and fifth thing happened. In my mind they happened almost in the same moment. In reality they could have been separated by weeks or months, but the impact was the same regardless. Deep discernment, prayer, and quiet time to pay attention and listen became invaluable."

Matthew read two more transformational books, one of which was *Man's Search for Meaning* by Victor Frankl, an Austrian psychiatrist and Holocaust survivor. Frankl wrote of how he was forced to assess whether his life still had any meaning under the horrific conditions he experienced in a concentration camp. He said that we all worry no matter our circumstances, and people worry no more or no less when in a concentration camp. It's not physically possible. He writes, "Worry is like a gas. It fills up every cubic inch of every space you introduce it into, no matter how much you introduce."

Those words had a profound impact on Matthew. "I realized that my worries about growing the company were mostly silly," he said. "I would worry no more or no less about making a $10,000 payroll as I would making a $100,000 payroll. Nor would I worry any more or any less than the guy living paycheck to paycheck."

Then Matthew heard a speech by a successful businessman at his business-owner group that piggybacked off his understanding of worry. "I don't work any harder than any of you guys," the businessman said. "My work just has more zeros behind it."

"I realized that I was going to worry no matter the size of my business, and I was going to work no less or no harder based on the size of the business," Matthew surmised. "The decision to worry or not worry, work more or less, was up to me no matter how big or small the business."

Completing the trifecta of life-changing books that Matthew read was *Rags to Riches* by Dick Farmer, the founder of Cintas. The book came from his brother-in-law Jeff Hite, who worked at Cintas. "I gave that book to him after I had learned what Cintas meant to him," Jeff said. "It was a common bond."[102]

Dick Farmer took over his dad and grandfather's shop-rag and uniform business and through many acquisitions transformed it into the largest business of its kind in the world. Farmer led the charge for that industry consolidation, and Matthew had the confidence that Nix Companies could do the same for the welding and fabrication industry.

"The book clearly goes through exactly what Matthew was going through, trying to talk his dad into taking more risks," Jeff said. "Matthew saw that the laundry industry was similar to the welding industry, but I didn't understand what he was saying. Then he explained that it was such a fragmented business and people want out. The mom-and-pop shops need an exit strategy. When he told me of his idea to buy all of these shops, I said, 'Dude, that's revolutionary.' No one had looked at it that way. But it was all because that's what Dick thought he would do with the laundry service."

As Matthew read *Rags to Riches*, the parallels between the early days of Cintas and those of Nix Companies became so obvious that the hair on the back of his neck stood up. At that moment, he had the conviction to go for it.

## CHAPTER 23

# Setting the Plan in Motion

Matthew knew there were several tasks he needed to complete to set his plan in motion. First, he vetted his idea with mentors. "I talked about my idea with any businessperson I respected who would listen," he said. He received much encouragement, some skepticism, and several suggestions that made the plan better.

"Allowing or forcing myself to talk through my ideas was as powerful as the advice received," Matthew said. "Often, before I finished my statement, I knew that a part of my idea needed work. Verbally articulating a concept can bring either clarity or murkiness."

Financing was the next step. Matthew explained:

In any business, there are three options for financing growth. Fund it through operating profits (cash flow), equity (investors), or debt. Typically, most companies try to fund their organic growth through cash flow and will occasionally utilize debt for larger purchases. This was the route we had taken until this point. The scale of my new plan obviously ruled out financing it through cash flow, so that left two other options and a difficult philosophical question. Do we want outside investors?

Our first experience with an investor, JZ, had gone well, but we had since bought him out, and the idea of keeping the company closely held was appealing. Especially since we are a fifth-generation family business hoping to continue to the sixth. However, we had to strongly consider the idea of bringing on a partner or partners to help fund growth. We wanted to maintain control of the company as the majority shareholder. Many businesses go this route and experience tremendous expansion and success. One party manages the day-to-day business, and the other party funds the growth and provides strategic direction and oversight. It's not a bad business model, especially if you have a good partner and both parties have the same values and long-term expectations. We had our share of courters for the partnership, but after a lot of soul-searching, we decided that wasn't the route we wanted. At least not for now. So that left debt.

Matthew needed a strong banking partner that would stick with Nix Companies for the long haul and meet his ambitions for growth. The company was still banking with the local Community State Bank. Matthew's plan surpassed the local bank's lending limit, so he needed to look elsewhere. After interviewing several bankers, he and the team settled on German American Bank (NASDAQ: GABC), a midsize institution headquartered in Jasper, Indiana. Like Nix Companies hoped to do, it had grown significantly through acquisitions. In 2023, Matthew would join the bank's regional board of advisers.

"I liked that they felt like a blend between the smaller-relationship bank I was used to and a larger institution," Matthew explained. "Their slogan is 'Large enough to serve you and small enough to know you' [foreshadowing the key slogan that in a few years would define Nix Companies—"Be Big, Act Small"]. I also had a mentoring friendship with Gene Pfeiffer, a former member of the bank's board of directors. With a bank apprised of my business plans and a much larger lending limit, we were ready to move to the next step."

Matthew needed an attorney. He worked with a local estate-planning lawyer who had met his past legal needs, but Matthew began to ask him increasingly complex questions. The lawyer told Matthew it was time to find a corporate attorney and connected him with a sharp, up-and-coming lawyer in his firm. Dan Robinson was about Matthew's age and had a young family. The two clicked immediately.

"Over a beer, I shared with him my ambitions for growth and my limited resources at that time," Matthew said. "I told him I wanted a lawyer who would be with us for the long haul. Not someone who was going to nickel-and-dime us to death while we tried to get our feet under us. He appreciated where I was coming from. After all, he was in a similar boat. From that day forward, Dan has represented Nix Companies as our corporate attorney."

A lawyer who understands a business intimately can give better advice. Matthew repeatedly tells his team and those he mentors, "You can ask the smartest businessperson you know [or lawyer in this case] for advice, but if they have no context or situational awareness, it's worthless advice."

Many lawyers are apt to say no to ideas that are revolutionary. Matthew bristles at that mode of operation. "If you aren't careful, a lawyer can keep you from doing anything new. In my mind, the lawyer's job is to help the client fully understand risks, provide strategies to mitigate those risks, and give other options regarding a plan. Their job is not to protect a client from every risk. The only way to accomplish that is to hide and never commit to any business transactions. That's not reality. Dan strikes a balance between helping us take risks and offering options to mitigate them."

The next step was to be proactive and ensure the family was on board. Matthew called a family meeting in a conference room at the chamber of commerce in Evansville. Bill and Donna had sold out to Matthew and Adam but were included because they were financing the buyout to their sons over ten years. Lindsey and Adam's wife, Lacey, also attended. This was the first time Matthew talked with the family about his goal to focus on acquisitions.

"I started the meeting by unveiling my vision statement, how it came to be, and how I planned to achieve that vision with my growth strategy," Matthew recalled. "I then laid out some long-term goals: big, hairy, audacious goals—BHAGs, pronounced 'bee-hags,' a term taken from the book *Good to Great*."

Author Jim Collins states that a "BHAG serves as a unifying focal point of effort, galvanizing people and creating team spirit as people strive toward a finish line. Like the 1960s NASA moon mission, a BHAG captures the imagination and grabs people in the gut."[103]

As a young worker at the welding shop way back when, Matthew had likewise learned from his father that to develop a strategic plan, one must start with the end in mind. Bill said, "Son, if you can't see the end of a project before you start, you are screwed."

To cap off the meeting, Matthew brought in a guest speaker, Kevin Koch—CEO of Koch Enterprises, the largest family business in Evansville, which grosses more than $1 billion annually. Koch talked about his company's history, his family's growth journey, and the corporate structure that enables them to remain family owned while operating at a magnificent scale. Last, he shared the good that his company and his family are able to do for the community because of the business.

"By the time the meeting concluded, I had set the tone for where we were headed," Matthew remembered. "The challenges that lay ahead and the risks we were taking were candidly clear. I felt like we *had to* go for it. Although there was no formal vote or a 'Let's do it!' chant at the end of the meeting, ultimately everyone supported me. There was a calm, matter-of-fact acceptance of the path we were about to embark upon."

At the family meeting, Matthew expressed his intention to establish an executive board of advisers, consisting of five businesspeople from outside the company. After an initial onboarding to bring them up to speed, they would meet quarterly with the leadership team to review financials and key performance indicators and discuss growth strategies. The advisers would approve the annual operating plan, all major transactions, and the executive-level compensation plan, including Matthew's.

"This oversight was meant to give my parents and our minority shareholders transparency and comfort, as well as provide guidance for myself and our executive team," Matthew explained. "It has been the single most important management implementation we have made. The board has guided us through many challenges and offered insights, allowing us to prosper beyond what we might have on our own. Members provide a sounding board when we vet opportunities, particularly acquisitions."

With the foundational work in place, in 2016 Matthew called the first annual off-site senior-leadership team meeting and officially launched the grand vision.

During the meeting, the company's initial five-year plan was born, leading to explosive growth from twenty employees to one hundred team members.

To cap off the day, Matthew brought in guest speaker Ira Boots, the CEO of Berry Plastics, a global plastics company headquartered in Evansville. Ira and his team had developed their first long-term—hundred-year—plan and vision to be the number one plastics packaging company in the world. They achieved the vision much sooner, in only about thirty years. Ira led the organization through many acquisitions and eventually its initial public offering on the New York Stock Exchange.

"He told us the limit to Nix Companies' growth was the limit of our imagination," Matthew recalled. "He made us believe that we had it within us to do just what he and his team had done. The outcome of that meeting was the development of a strategic plan on one sheet of paper."

The paper included the company vision and long-term "big, hairy, audacious goals" with five- and thirty-year timelines. The primary five-year goal was to become the number one metal fabrication company in the tristate and to reach $10 million in revenue. "Number one" would be measured in terms of having the most market shares and the best team members among their competitors. In just three years, at the end of 2018, Nix Companies achieved that five-year goal. The leadership group grew to thirteen. Ira Boots was invited back to celebrate the achievement he had helped to inspire.

Also during that senior-leadership meeting in 2016, Adam uttered words that would become the company mantra. The team sat around a table in a brainstorming session, with each member writing down ideas. They took turns reading aloud what they had scribbled. When it was Adam's turn, he said, "I only wrote down four words: Be Big, Act Small."

"None of us said, 'Aha, that's our mission statement,'" Matthew recalled. "But every one of us knew Adam's words were going to be important."

Also that year, Nix Companies won several awards, including Posey County Business of the Year from the Posey County Southwest Indiana Chamber, Indiana Companies to Watch from the Indiana Small Business Development Center, and an Inc. 5000 Award from *Inc.* magazine, recognizing them as the twenty-seventh fastest-growing manufacturing company in the US and the twenty-fourth fastest-growing company in Indiana, overall. At this writing, *Inc.* has named Nix Industrial one of the fastest-revenue-growing companies in the country for seven consecutive years, and NIX has been named one of the best places to work in Indiana four times.

For the next couple of years, the "Be Big, Act Small" phrase circulated and was utilized in various ways throughout the organization. One day, Matthew and Adam Schmitt, vice president of business development and administration, were creating the company's first version of the "Vision, Mission, Values" pocket cards that all employees receive when they join the team, and Adam's four words gained greater meaning.

"At the time, our mission statement was a paragraph [not uncommon for many organizations]," Matthew said. "No one could remember it, and that frustrated me. As I tried to decide how to make it fit on a wallet-size card, I looked over at a piece of paper bearing the company slogan. I grabbed the phone, called Schmitt's office, and said, 'Be Big, Act Small'! It's our frickin' mission statement!"

At that moment, the words Adam Nix had scribbled in a notebook years prior took on a life of their own, and a new, concise mission statement was launched.

The beauty of the statement is that it brings together two opposing thoughts. "The bigger we get," Matthew said, "the harder it is to act small and the more important the statement becomes. The bigger we get, the more we need the mission statement as a reminder to stay true to our roots."

**CHAPTER 24**

# The Value of Underutilized Potential

With all the "seats on the bus filled and the right people in the right seats," as Jim Collins writes in *Good to Great*, it was finally time for Matthew and the executive team to put the plan into action.

Early in 2017, Matthew had a chance encounter with Joe Kratochvil, a consultant with Clear Stone Solutions. Joe asked whether Matthew wanted to be involved in a business he was starting. After listening to his pitch, Matthew respectfully declined the offer, but he had another idea about how the two could work together. Matthew asked Joe if he would like to represent NIX in the search for acquisitions. Joe had done similar work in the past, and he accepted immediately. He would search out deals anonymously, allowing NIX to approach competitors without them knowing in the early stages.

In late June, Joe learned of the opportunity to acquire Superior Fabrication in Rockport, Indiana, about one hour from Poseyville. Owned by founder Curtis Drake, from Owensboro, Kentucky, Superior Fabrication had provided structural steel fabrication to commercial construction and heavy industrial markets for twenty years. At the time, it seemed like a long way away from Poseyville, but the NIX leadership was open to exploring it.

"While I was on a vacation with Lindsey's family in Hilton Head, South Carolina, I received the nondisclosure agreement to sign," Matthew recalled. "Within a few hours of signing, I had the prospectus and realized for the first time that this was the real deal."

Matthew was interested in structural steel fabrication because at that time NIX was not set up well for this type of work and struggled to compete. They had just lost a $500,000 job to a local competitor. Had NIX owned Superior, they likely could have won the project and turned a profit.

The potential acquisition process began with a visit to Superior's operations. Ryan Weinzapfel, NIX engineering manager, accompanied Matthew on the first trip. Later the entire executive team paid a visit. Matthew and the leadership team were eager to invest in Superior because of the company's technology, both in the shop and in the office.

"Superior would get us more involved with commercial construction and capital projects," Matthew said. "At this point, most of our fabrication work was more maintenance or operations related within the industrial fabrication space. Diversifying into capital projects would allow us to benefit from economic upswings and grow our revenues in ways we could not achieve otherwise. Superior Fabrication had a nice facility, as well as a CNC [computer numerical control] beamline consisting of a saw, drill, and coping robot."

The facility had underutilized potential. It was what Matthew called "operationally sound and sales distressed," meaning a business has all the infrastructure, systems, and processes in place to grow, but they need more sales and the associated labor to execute the work. The team would later realize that any company with underutilized potential was the ideal candidate for an acquisition.

"This sort of scenario banks on our core competencies," Matthew explained. "We know how to increase revenue and attract talent, proficiencies crucial to growth that often elude the best owners who come from blue-collar backgrounds and tend to focus more on operational components while neglecting what fuels the real economics of their business. Superior had all the makings of that model, but at the time we didn't *know* that was our model. We just knew intuitively that Curtis, who was willing to stay on board, along with his shop, systems, and technology, made for a perfect fit for a partnership."

Matthew reported the opportunity to the NIX board of advisers and received their blessing to move forward. The advisory board saw the great need for growth in order to afford the top-notch executive team already in place—an expensive burden on the small company that NIX was then.

"I viewed these key hires as I would any crucial asset," Matthew said. "It was an investment that might take a few years to fully return. This strategy is fine as long as you don't run out of cash, and in 2017, the management overhead was burning its share of cash. We knew we needed to be proactive about finding a way to grow, and this seemed like the best opportunity we had, so I forged ahead."

One August evening, Matthew and Lindsey invited Curtis Drake and his wife, Charlotte, a couple in their early fifties, to dinner at a restaurant in Newburgh, Indiana, along the Ohio River. Because Curtis had offered to continue working for an extended period after the acquisition, Matthew knew their relationship would be critical to the deal. It was important for their families to mingle and gain a sense of comfort with one another.

"We had a wonderful dinner together," Matthew said. "By the end of the evening, as we departed, Charlotte and Lindsey shared a hug. I knew at that moment we were going to make a deal. I just didn't know how hard of a journey it was going to be to get there."

To complete preliminary negotiations and hand-deliver an offer, the two parties met at a neutral site in downtown Evansville. NIX attendees were Matthew; Lindsey; Jared Bael, vice president of finance; and Joe Kratochvil, mergers and acquisitions consultant. Representing Superior were Curtis and Charlotte Drake; Richard Clements, mergers and acquisitions consultant; and Brian Schulte, their certified public accountant. The NIX party arrived early to prep. Later, Matthew deemed the early arrival "a stroke of luck . . . or divinity."

"I had just listened to a podcast about negotiations," Matthew explained. "I learned of a conference table strategy. Rather than have the two parties sit across from one another, which creates an adversarial psychology, the parties should sit in every other seat."

It was a subtle but brilliant tactic. Before the Superior Fabrication representatives arrived, Matthew positioned NIX people around the table in every other seat.

"The Superior people had no option but to sit 'among us' instead of across from us," Matthew said. "I put Lindsey and myself on one side of the table with seats saved for the Drakes and placed Jared and Joe on the other side with seats saved for Richard and Brian. This further sent the signal that we [the buyer and seller] wanted to make a deal. They [the council] were there to help that happen. Too often accountants, lawyers, and advisers can get in the way of a deal. I wanted to ensure that didn't happen and that we could play 'good cop, bad cop.'"

The Superior people arrived, and Matthew stood as they entered, politely motioning for Curtis to sit beside him as if to say, "Hey, old buddy, I saved you a seat." He knew Charlotte would follow suit.

They reviewed the NIX offer and had questions and counters for each section.

"Soon, I could tell we were so far apart that there was no way we could possibly reach a compromise," Matthew remembered. "I said, 'Look, we would love to own Superior. Curtis and Charlotte, you have built a really nice business, and Lindsey and I really love you guys as people. I wish we could reach an agreement, but I don't see any possible way we are going to meet in the middle.'"

Relying on the seating arrangement, Matthew continued, "I think we both want to make a deal, but if these guys"—pointing to the other side of the table—"can't figure it out, I don't want to waste your time. I respect you guys too much, and I don't want to waste our time either. We are just too far apart."

Matthew began packing up his papers to leave. He truly was ready to walk, but he was hoping they would back down. Then the seating plan kicked into action. On cue, Joe turned to the other men on "his side" and said, "What do you all think about the Nix and the Drake parties leaving the room for a bit, and the four of us hash things out?"

They nodded in approval, and both Matthew and Curtis agreed to the idea. The two couples walked next door to a restaurant and chatted over Diet Cokes. They did not talk about the transaction except to say that they hoped their respective advisers could find a way to get the deal done. After less than an hour, Matthew received a text from Jared outlining the compromise the two parties had reached. On Halloween 2017, both parties signed the letter of intent.

"We still had to go through due diligence and negotiate the intricate details, such as working capital, legal responsibilities and liabilities, and how compensation plans would be calculated," Matthew recalled. "The month of November was brutal. As the accountants and lawyers got involved, the details were overwhelming."

Meanwhile, Matthew and the team worked as if the deal were going to close at any minute; everything had to be ready to hit the ground running. "Juggling this was emotionally and physically exhausting," Matthew said. "At times I didn't even know what day it was. It was one of the most stressful yet exhilarating times of my career. I went to the office at 6 a.m. and traded emails and text messages with Jared until midnight. We were working on the deal seven days a week."

For Jared, who had just started working at NIX in August, the month Matthew and Lindsey first met the Drakes for dinner in Newburgh, the experience was baptism by fire. No wonder Matthew claims that the craziest thing Jared ever did was join Nix Companies.

Emotions ran high, and at one point, Matthew began ranting in a text message with Jared and Joe, calling Brian, the CPA for Superior, an "ass."

"As soon as I hit send, I realized I had accidentally replied to a text that included Richard, Superior's mergers and acquisitions consultant. I had bashed the opposing counsel with someone from the other side of the deal on the text thread. I was immediately sick to my stomach. Within thirty seconds, Joe called me. 'What the f--k dude?!' After conversing for a few minutes, we decided to just own it."

Joe called Richard and spun the incident as if Matthew had

intended to include Richard in the text. He spoke to him adviser-to-adviser. He said, "Look, Matthew is obviously emotionally attached to this deal. I've counseled him and told him he shouldn't say things like that, but I think it's a good sign that he trusts you enough to say things like that in front of you. He obviously trusts the two of us to get this deal done. We need to work for our respective sides and do it."

He easily convinced Richard it would not do anyone any good for Matthew's comment to go farther than that group. This was an excruciating lesson for Matthew, who was thirty-two at the time. "To this day, I can still feel the pain. Now I reread any important text a couple of times and look over the distribution list before I hit send. The irony to the story is that Brian and I are now friendly. I have played a couple of rounds of golf with him. He's a good guy. We can laugh about those things now, but it wasn't funny at the time."

With so much invested emotionally in the deal, Matthew couldn't bear the idea of it falling apart, but in late November, he was prepared to walk away. The two parties were at an impasse on some financial details. If the Superior people didn't come over to the NIX point of view, Matthew didn't think he could get the deal financed. Then, unfortunately, more inappropriate words flew. This time they came from mild-mannered Jared.

During a phone conversation, Jared, customarily cool as a cucumber, lost his decorum and cussed out Brian. In turn, Brain spat, "If that's the way you're going to be, I am not sure I want my client working with you!"

Jared immediately called Matthew, his new boss but longtime friend, with his tail tucked, worried he had screwed up royally. Matthew had been where he was just a few weeks before.

"I knew the sinking feeling that tells us we have crossed the line," Matthew said. "It's meant to remind us not to do that again. Jared is a fair, polite, ethical, and levelheaded guy, but he's also a fierce competitor, and when he feels threatened, there isn't an ounce of flight in that guy. It's all *fight*. I appreciate that about him. He doesn't get worked up often, but when he does, he's tough. I calmed him down and told him I would call Curtis to smooth things over."

Brian had already called Curtis about Jared's unseemly behavior. Matthew explained to Curtis that if he chose to join NIX, there would be no more occurrences like this one. Off the cuff, Matthew added, "By the way, once you join our team, isn't that the guy you want in your corner fighting for you?" Curtis chuckled and agreed. The crisis was averted.

After nearly two weeks of back-and-forth between the two counsels and legal bills racking up, Curtis called Matthew with a brilliant request. He asked if he and Charlotte could meet with Matthew and Lindsey to review the contract together. No lawyers or advisers present. Matthew quickly obliged, and the two couples met on a Sunday afternoon at the Nix home. Lindsey made a pot of soup, and they all ate lunch, along with the Nixes' two sons. Matthew and Lindsey purposely did not hire a babysitter.

"I wanted Curtis and Charlotte to be in our home as part of our family," Matthew said. "After several hard-fought weeks, the atmosphere was warm and friendly. We sat together and went through the red-line copies of the legal documents. We laughed at ourselves, a couple of welders trying to interpret legal jargon. We looked some things up on Google. We compromised on many items. When we were done, we had a list of a few matters that we didn't understand enough to make a call on. We decided if we didn't understand their purpose, they couldn't be that important."

At the Nix kitchen table, the four reached a deal. The Drakes had put everything on the line and built a nice company. Matthew and Lindsey had put everything on the line to be in the position to buy such a company. They shook hands and vowed to not let the lawyers "screw this up." The next day, they each called their respective lawyer and told him he had twenty-four hours to reach an agreement with the other lawyer or they would both be fired. The attorneys came through, and an agreement was reached. Superior Fabrication was acquired in January 2018.

"Matthew and I really got to know Curtis and Charlotte Drake," Lindsey said. "We consider them good friends. We brought them into the family. We don't just *say* we're a family business; we really bring them into the family. That's important to us."[104]

In a press release announcing the acquisition, Curtis was quoted: "Since I started Superior Fabrication in 1997, I have put my heart and soul into every project that went through the doors. I feel comfortable knowing that we are turning Superior Fabrication over to a company that will do the same. I wasn't out there shopping for someone to buy my business, but when they came along, and we got to know them and they got to know us, we knew that this is God's plan. . . . I'm excited to have the opportunity to participate in what the next generation will do to grow and expand what I started."[105]

The first day after Nix Companies took over Superior, Matthew, Adam, and Lindsey met with each employee one-on-one. It quickly became clear that significant cultural challenges lay ahead. "In one meeting, the painter actually admitted to being a drug dealer and told us if things didn't work out with us, he would just go back to 'slanging that dope,'" Matthew recalled. The three did their best to keep their composure as they sat in front of the admitted drug dealer. As soon as he left the room, they looked at each other as if to say, "Did that just happen?!"

With trepidation as to the employee's response, they parted ways with him in short order. "It was nerve racking, not knowing how someone like that would react," Matthew said. "It's an eerie feeling to be looking over your shoulder in these situations or worrying about the safety of your staff and family."

"We weren't used to any of that," Lindsey recalled. "Curtis told us that he had some guys with history, but they were good guys, and he tried to keep out the riffraff. But that was one type of riffraff that was new to us on the first day. Superior has an awesome crew now, and they are rock-and-rolling, having a record year."[106]

While this was the most extreme example of the culture clash between NIX and Superior, it was not the only one. Within the first ninety days, more than half of the Superior staff turned over, and by the end of the first year, only two of the eight people who had been there when NIX took over the helm remained on the shop floor.

Matthew said,

> The two who stuck it out wore ankle-monitoring bracelets but turned out to be exceptional team members. We helped one obtain his associate's degree by accommodating his schedule with flexible work hours. He left his criminal history behind and bettered his life. We were proud of him and happy to help. He left on good terms to a great career in maintenance for a local municipality. The other remaining Superior team member also transformed his former life as a criminal and became our shop foreman. We are proud of what he has accomplished both personally and professionally. Today, he leads a dozen or more people on the shop floor of our structural division.

Angela Kirlin, director of HR, described NIX as a "second-chance company," willing to give a leg up to someone who has had a hard life or made poor choices and wants to better their lot. One month after she joined the company in February 2018, Superior was acquired. Beginning on her first day of work, she was thrown into the task of culturally integrating the company and filling several open positions. In her first year, she hired about fifty team members.

Perhaps it was the Superior acquisition that introduced the NIX team to the idea of being a second-chance employer. Matthew said, "Curtis has a heart for guys who need a second or third chance, and I respect that about him. In at least one instance, the second chance completely changed an individual's life and led to a generational shift. If Curtis had given up on this person, no telling where he'd be today."

"I was trying to embed the brand-new location with the NIX culture that I had not even experienced yet," Angela remembered. "It was overwhelming. The funny thing is, I'm not from this industry. I don't know about welding. But I know how to welcome people."[107]

Angela slipped on her work boots and went to work next to the people on the shop floor, job-shadowing each division to obtain a sense of their daily experiences while also learning the terminology of the business. "I tried to create a family environment, and that really helped. I didn't know these people, but I wanted to welcome them into this family because I knew Matthew's values, and that's what is really important to me."[108]

Prior to the big downturn of the structural steel business in 2009, Superior was producing as much as $8 million a year in revenue. After the downturn, the business yearly declined to as little as $2 million. Unfortunately, Curtis was forced to downsize his staff. "The sacrifices he made to keep the business afloat were impressive and inspiring," Matthew said. Short on staff, Curtis had to jump in. He handled the estimating and helped with detailing and project management while supporting the shop floor.

Matthew explained,

> When we took over, we provided the support to allow Curtis to do what he's great at—run the operations. We took on recruiting and staffing, moved the accounting to our main office in Poseyville, and immediately got to work on adding revenue. We hired an exceptional estimator who made an immediate impact in growing the business. Additionally, our outside sales and marketing team brought in new business opportunities. Within the first year, we had nearly doubled the revenue, and we continued the trend for the following two years.
>
> A big contributor was the expansion of the business into adjacent markets. As an example, in 2021, Lindsey's sister Mandy Crislip, who joined the company part time to help expand into the Indianapolis market, landed the company a major project in downtown Indianapolis. The Bottle Works

Hotel was a renovation of an old Coca-Cola bottling plant that was converted into a high-end hotel. This project really put us on the map as a major player outside of our local area In 2023, this business unit did over twenty-two million in annual revenue. Having a strong operational leader such as Curtis along with our support services was like pouring fuel on an existing fire.

Looking back, Matthew said, "The shop floor metamorphosed. The layout did not change, but the atmosphere and culture did—from the tangible, like all-new LED lighting, cleaning, and organizing, to the less tangible: the more valuable change in morale. Transforming the team and the culture of our first acquisition remains at the top of the list of our most astounding and beneficial accomplishments."

Curtis Drake was the first recipient of the annual Charles Henry Nix Entreleadership Award. "I consider him not only a team member but a friend and mentor," Matthew said. "It is an honor to carry his Superior Fabrication legacy forward."

## CHAPTER 25

# Return to Poseyville

### HUMOR AND HEAVY INDUSTRIAL SALES

More community friends returned to their roots to work for Nix Companies in peaceful Poseyville after working in the big corporate sphere. When Corey Wilzbacher left Kaman Industrial Technologies in 2016, people thought he was a fool. Little did they know, he was on a personal quest.

"Everyone said I was crazy for going to work for Matthew in a family-owned business," Corey said, reminiscing with a laugh. "Everyone thinks I'm nuts anyway. So whenever I did this, it was like *Oh, my God, here he goes again.*"

A tall, burly, animated man, an open book who shoots from the hip with humor, Corey had been a troubleshooter for heavy industrial equipment while building customer relationships at Kaman. "I tried to make things more reliable for people. I left Kaman because of the big-business world."

Corey was originally from nearby Haubstadt, where his family farmed, and Corey's grandfather had brought farm equipment to Sonny Nix for repair, so, like many NIX employees, Corey grew up visiting the welding shop. Lindsey grew up kitty-corner from Corey.

"I knew the Nix family," Corey said. "They were good, honest, hardworking people. I knew I wanted more of an intimate feel, a place where I could have a conversation with a customer and we could get to a right answer, whatever that was. No limitations."

Corey was volunteering, cooking in the kitchen at Saints Peter and Paul Catholic Church in Haubstadt, when he saw Brandon Wright, who encouraged him to talk to Matthew. Corey reached out to Matthew, and the two talked for a few months about his interest in a possible change. "I'd been hearing of NIX for years," Corey recalled. "I drove by and I saw the yachts being built. I thought that was cool, and I was super proud of Matthew. But we never had a conversation about it. It's just neat whenever you see something and you understand the work that it takes to get there. He was just grinding. I looked at it from a distance, and I admired it."

When he was hired, Corey left perks and stability at Kaman, such as good pay and eight weeks' vacation. His father asked, "What are you doing?"

"I've got to do this for peace of mind," Corey explained to him. "And I'm going to take less money to go do it because I believe in Matthew and his idea. I feel like I can help that family go."

Looking back in 2023, Corey said, "My dad still thinks I'm crazy, but now he sees that with some drive and some freedom, you can do about anything you want to. It didn't matter if we made two cents on our dollar or eighty cents on our dollar. We put in the same effort, and I have a team and brothers and we just figure it out. I didn't walk into a whole book of policies and procedures to cover everybody in an umbrella. I became a part of the freedom for us to create the playbook—the standards for our business. We have fun and work."

Corey came on board to do sales, utilizing his relationships and knowledge accrued from past experiences. He eventually brought larger projects to NIX. "It took eight months for us to get a purchase order over a hundred thousand dollars," Corey said. "When that happened, it was like we had hit the lottery. Our ship had sailed. Because most of the time, back then, it was five-thousand or twenty-thousand-dollar jobs."

When Corey began working with power plants and large companies like Duke Energy, now one of Nix Companies' main customers, it gave NIX some freedom and cash to grow. "Basically we built a whole new division outside of taking care of the farmers and commercial market share. When I came, I added another bucket with heavy industrial—power plants, steel smelters, aluminum smelters. Now we're in every power plant in the tristate area. And also down at Florida Power and Light, a nuclear power plant, and many others."

The NIX way of doing business appeals to customers in numerous ways. For one, the team does what they say they are going to do. "And if we can't, we notify them right away," Corey said. "Not three weeks later. Because we're not afraid of losing the money. We're afraid of losing the relationship. We're not trying to just bid low. If they want a cheaper price, well, they gotta change the way they want it built."

Most of the NIX team grew up in the community, with the same disciplined upbringing. "Our parents raised us right," Corey said. "We're raising our children the same way. Not everybody gets a trophy. You have to work hard, and you can be honest, and this is what you can accomplish when that happens."

Some misunderstand Corey's description of NIX people. "I tell everybody NIX is just a bunch of misfits who found a home because the traditional definition of success is financial or material possessions. A lot of us took pay cuts to come to work for Matthew. It was for peace of mind. We're not the same as everybody else out there. We are a bunch of oddball misfits who want to work our asses off."

For Corey, fully understanding his decision to shed stability for fulfillment took a few years, however. This shift occurred in 2018, after Matthew spent a weekend at Cursillo, a Catholic men's retreat that "helps people discover and fulfill their personal vocations while promoting groups of Christians who leaven their environments with the Gospel."[109] The Diocese of Evansville website states, "The purpose of the Cursillo Movement is to make Christian Communities possible in neighborhoods, parishes, work situations, and other places where people live the greater part of their lives. The Cursillo makes it possible for anyone in the world to live a Christian life in a natural way."[110]

"Matthew came back from Cursillo on fire with his faith," Corey recalled. "I told him, 'I need what you got.' I left Kaman just for the hell of it, and you just don't do things like that. My father and all kinds of people had been in my ear, telling me that. I was on some sort of journey, trying to make myself happen. Figure out how I could be who I am. Through that whole shift, I thought maybe God's pushing me toward going to this Cursillo because I need to learn how to be a better leader. So, Matthew sponsored me to attend Cursillo. We connected over that. It's a strong weekend to evaluate yourself, understand the changes you need to make, and then live your life that way, day-to-day."

Corey's Catholic faith and the Cursillo experience have added immeasurably to his kinship with the Nix family, their work ethic, and their humbleness and quietness of good deeds. "Matthew and I are on a faith journey together," Corey said. "So the connection is more than just employee and owner. It's spiritual for me. There's a lot of faith in our company. God put Matthew in the CEO role, and God put each one of us in each other's lives at the right time. When we could manage it and when we were ready for it."

That said, Matthew never wants to come across as imposing his beliefs on others and doesn't want anyone to get the impression that they must be Catholic to be in a leadership role or work at NIX.

"We just treat people with love, dignity, and respect," he said. "It's pretty simple. We have our convictions and will not waver in what we believe or value, but we are open and respectful to all people from any faith background, ethnicity, or lifestyle. We welcome anyone qualified for a job. I think diversity is healthy for our business, and I don't mean just checking boxes or hanging up DEI [diversity, equity, and inclusion] banners. It's about bringing in people from different backgrounds, different ideas, and fresh perspectives."

Soon, a hometown boy, a fixture in the lives of the Nix family, would be ready to become CFO of Nix Companies.

• • •

## HIRING THE "MAYOR OF POSEYVILLE"

When an acquisition fell through in early 2017, Matthew became restless. Now that he had his plan, he was ready to make big moves. However, two puzzle pieces needed to be in place to complete the NIX picture: a controller/CFO and a human resources manager.

Lindsey still managed the company's finances and performed some HR tasks. Other members of the management team took up the slack on HR needs, but no one handled human resources full-time in a manner that could move the company forward at greater speed.

Lindsey was stretched thin and did not have time to devote to marketing, the job she first came on board to do. She often worked from 7:30 a.m. to 5 p.m. in the office, picked up her two diaper-clad sons (Mason Eugene was born in 2016) from daycare, cooked supper, did dishes, laundry, and more, and then put the kids to bed before going back to work on her laptop until midnight.

"I was working constantly at that time too, so I wasn't much help to her," Matthew admitted. "To make matters worse, her role did not facilitate a flexible schedule. When you are the person who's responsible for paying the bills and running payroll, the work never stops. While Lindsey had done an extraordinary job of getting us to that point, we both wanted to bring on someone with a corporate finance background who could help lead our company's finances for the next thirty years."

Matthew joked that the CEO and the CFO should not live in the same household. "It was my job to spend the money and Lindsey's job to save it. In hindsight it was a great balance," Matthew said with a laugh. "She was perfect at keeping me humble and grounded. I would come home excited about a big job. 'Honey, I just sold a million-dollar job today. I need to buy another piece of equipment!' She would say, 'Great. How we gonna pay for it? Oh, and please take out the trash.'"[111]

The couple lived that life for about five years.

"We were doing another acquisition, and I cried uncle," Lindsey recalled in 2023. "I said we need some help. It had been utter chaos during the high-growth years. Today, we are up to one hundred sixty employees, with about one-third of those attained through acquisitions. Hiring the first fifty employees was much more difficult than the next one hundred. Looking back, those early years are like a fog."[112]

The team wanted a smart, humble, and hardworking CFO who fit the NIX culture. And, of course, that someone had to be trustworthy. Lindsey set the bar high when she said, "I need that person to be someone I would trust to care for my kids. We're basically handing over our checkbook to a stranger."

Matthew had someone in mind, and his intuition led to déjà vu. Just as he had correctly predicted to his brother five years earlier that Brian Merkley would one day work at NIX, Matthew said, "Jared Baehl is going to be our CFO someday."

The Nix brothers had known Jared for their entire lives. Born and raised in Poseyville, he frequented the welding shop with his father, an electrician who sometimes did work for Sonny and Bill. Jared grew up hunting with Bill and Matthew, who was like an older cousin.

However, even more significantly, Jared is Adam's lifelong best friend. They grew up playing baseball together with their fathers coaching, from Little League to high school, where Adam as catcher and Jared as pitcher led the North Posey team to win two state championships. Adam was the best man at Jared's wedding and is the godfather of Jared's daughter.

When Matthew made his prediction, Jared worked for a public accounting firm in Evansville and owned a house on Poseyville's Main Street, where he and his wife were raising their children. He was president of the youth baseball league and head of finance on the parish council of the church that he and the Nix brothers attended growing up. The connections and bonds between the three men ran deep.

Adam's response to Matthew's declaration also mirrored the Brian prediction. "We can't afford Jared. Besides, he won't want to come here. He's going to be a partner someday where he is."

"Adam was right on both counts," Matthew recalled. "But I didn't care."

Matthew looked at his brother and spoke the words handed down by their father: "Well, you just hide and watch."

The brothers and Brian Merkley, vice president of operations who also grew up with Jared and the Nix brothers and played football with Jared and Adam, invited Jared to play a round of golf in the fall of 2016. About three holes in, Matthew casually said, "We're going to be looking for a controller soon."

"I thought that might be why you invited me out here," Jared replied with a smile.

The men laughed, and for the remainder of the day, Matthew presented his best sales pitch on why Jared should join the company.

Because the leadership team was struggling to justify the expense of a full-time controller, Matthew and Adam talked to Jared about coming on board to handle operational management with Nix Equipment, formerly Tri-County Equipment, in addition to taking over the company's finances. A few weeks later, after giving it much consideration, Jared declined the offer.

Looking back, Jared said, "I don't know if the position wasn't right or if it just wasn't the exact right time. I simply wasn't sure about it. It didn't feel right. I'm not a person who moves real quick on anything. I'm a processor and a thinker."[113]

Matthew had a feeling that this wasn't the end of the Jared question, but he was determined to move forward and hire a controller. In the summer of 2017, with the next acquisition drawing near, he got serious. The position for a vice president of finance was posted. It carried the responsibilities of a typical controller and a few responsibilities of a typical CFO. "I would maintain the other aspects of the CFO role," Matthew said. "The job description made clear that the person hired would be expected to transition to the full-time CFO role in the near future and, with company growth, lead a controller as a direct report."[114]

Matthew hoped that Jared would apply but did not approach him. The job opening was circulated among businesspeople Matthew knew and posted on social media. Matthew was confident that Jared would learn the position was available. However, after a few weeks, Jared had not applied. Brian made a point of bumping into him and asked whether he knew there was an opening. "I wasn't sure they would want me since I had declined previously," Jared explained later. Brian quickly corrected that assumption and encouraged Jared to apply. The interview process began.

All candidates were interviewed using the same questions and quantitative scoring system. The leadership team did their best to keep emotion out of the interview process. But there is no stopping intuition.

"When we finished with Jared's interview, I told Matthew that if I died tomorrow, I would trust him and Heather, his wife, to raise our kids," Lindsey recalled.[115] In other words, Lindsey trusted Jared to handle the money and their business like a protective father.

It came down to Jared and another candidate. Matthew had the two candidates interview with the NIX board of advisers, of which only one was from Poseyville, and Jared knew him. Although the other contender was already in a CFO role and had a stellar track record, the board was supportive of Jared as the first CFO of NIX. Matthew paraphrased their collective comment: "If you want to bring on investors or go public and grow this thing like a rocket ship to sell it, hire the first guy. If you want a solid, long-term growth model and to pass the company on to the sixth generation, then hire Jared."

Jared did not have the career experience that the other candidate had, but the board was confident he would acquire the expertise and be around for the long haul. "Jared was as close to family as one could get," Matthew said. "I didn't want anyone to think we hired him just because of our relationship. We jokingly refer to him as the 'mayor of Poseyville.'"

Poseyville doesn't have a mayor. It has a town council president—Jared's father, Michael.

Jared summed up his hiring and profound responsibility:

> I feel good that I interviewed just like anyone else. I wanted to come in and succeed and prove myself right from the get-go and show how I could contribute. I don't take my position

lightly. And no one else does either. All of us respect each other a ton, and I knew when I was hired what this position meant to everyone, to have confidence to make decisions based on good data. It's everyone's livelihoods. I think about the company and the family to make sure both are taken care of. That's my responsibility. The sugar on top is I do it in a place where I grew up. I care about Poseyville and the people in it. I get to work in a town that I enjoy living in for a company that's great to work for.[116]

No better words could be spoken by the "mayor of Poseyville."

• • •

## ENTER THE CINTAS CONNECTION

The last puzzle piece to snap into place for Matthew's growth plan was to hire someone to focus on human resources—specifically, recruiting and onboarding new team members. The operations managers had been performing these tasks, and it was taking more and more time and, as a result, becoming sloppy.

As 2017 drew to an end, NIX was on track to close on a major acquisition, so it was time to post a position for the company's first HR manager. "To our delight, we had a lot of really great candidates," Matthew said. "Just like our VP of finance position, there were three or four front-runners, and one of them was another hometown candidate."

Angela Kirlin was originally from the area, and she had worked at Cintas Corporation, the family-founded Fortune 500 business Matthew wanted Nix Companies to emulate. The business had held honesty and integrity as its central tenets since it started in 1929. Two years earlier, Matthew had pored over the book *Rags to Riches: How Corporate Culture Spawned a Great Company*, written by Cintas founder Richard T. Farmer, underlining sentences and making notes in the margins on how to expand and lead NIX.

Two sentences Matthew underscored were "Competitors can copy our sales material, our products and even some of our systems, but they cannot copy our culture. It is invisible to the outside world and therefore impossible to replicate."[117]

Angela, a positive, astute woman with dark hair, grew up on a farm outside of Poseyville, and her parents are close lifelong friends with Matthew's parents. Her brother and Matthew had been friends and teammates since they were children. The two families camped together in the summer. Angela graduated from Indiana State University with degrees in both business management and administration and obtained a master's degree in human resources training and development, then began her career with Cintas.

"Just like Jared Baehl, Angela was about as close to family as you can get," Matthew said. "On top of that, her résumé was solid. Angela had been the HR manager of Cintas in Louisville, Kentucky, where she was responsible for a few hundred people. This was another interesting turn of fate, since Cintas is such a part of our story.

"She had been following our journey on social media (like so many great team members we have brought on) and was really excited about the opportunity to apply to join our team in a role that she was seemingly made for," Matthew said.

Like so many other NIX employees, Angela grew up visiting the welding shop with her father, a farmer who needed equipment repair. "As a young kid, being around Bill and Sonny, I saw the mom-and-pop welding shop," Angela said. "And then I saw what Matthew was doing, his vision."[118]

Again, the NIX leadership team interviewed all the top candidates with an open mind, using a quantitative scoring system to rank them in the least subjective way possible. "Angela was the clear leader," Matthew recalled. "We felt her personality was perfect for what we were looking for. She was capable, and we were confident she would be loyal."

"I sat in on Angela's phone interview," Lindsey recalled. "She may not even know that. I didn't say a word. I just wanted to listen. Adam Schmitt ran the interview, and when he hung up, I said, 'She's it. I don't want to interview anyone else.' And she has been phenomenal."[119]

Although Angela enjoyed her time at Cintas and speaks of the company with admiration, eventually she felt the larger company wasn't where she belonged. So, when she had her son, she took a year off. "As much as I love him," she said, "I'm not a stay-at-home mom. I like interaction. I like the people side. I like helping people."[120]

She then worked for five years overseeing the GED (General Educational Development Test) program in eight counties near Jeffersonville, Indiana. "I saw adults with very rough upbringings obtain their high school diplomas."[121] After her second child was born, she and her husband built their home on the pasture at her parents' farm near Poseyville. Then, she saw the social media post for the NIX human resources position and applied.

"The culture is what drove me," Angela said. "I am a huge people person. HR was kind of a difficult choice because at Cintas, in the big corporate environment, it's a hard job. Very strict. In that plant there were about three hundred employees. So you have the good and the bad side. I like the small-town family feel at NIX, so I knew working there was the perfect opportunity for me."[122]

Angela jumped in full force and worked on the shop floor next to the welders. "They had a really fun time with me because I knew nothing," Angela laughed. "They were so welcoming and wanted to teach me what they do. Getting that one-on-one experience is a two-way street because it gave me an appreciation for what they do, but they also appreciated that somebody in administration, in the office, put a uniform on, laced up her boots, and got to work right next to them."[123]

## CHAPTER 26

# It's About the People

The NIX vision reads: "A world-class team; a nationally recognized brand; spanning generations."

A good vision statement conveys a clear and compelling dream for the future—a way for stakeholders to understand the meaning and purpose of a business. It's a summary of goals and why a company exists. "I knew our vision statement had to be something that was one hundred percent authentic to my beliefs and goals and something that I could really get behind," Matthew said. "I am the one who must 'sell' the vision to the team. If I do not believe it, no one will believe it."

In the beginning, Matthew focused on growth for growth's sake, and his ego was involved. He wanted to prove to himself that he could do it. As he matured with the business, he asked himself deeper questions: *Why am I doing what I'm doing? Is it all worth it? We had a perfectly good family business where we made a comfortable middle-class living. Why did we need to continue to push to grow?*

He wrestled with these questions for a couple of years, an intense period of introspection. He read books and spent time in reflection and prayer. "I kept asking myself, *If I'm not supposed to do this, then why did God make me this way*? I was envious of people like my dad and grandpa who led more simple lives. They had done very well by most worldly standards but were still godly and simple men. They were not restless like me. I wanted to be content. Finally, I recognized that we all have unique gifts, and my gift was to be an entrepreneur. To not harness and utilize my gift for good was to waste my talents. I did not want to be like the servant in Matthew 25:14 whose master gave him one talent and he only returned the one. I wanted to be the one who was given five and returned to his master with ten."

Even with that revelation, Matthew struggled with *how* to live out his talents. "Then one Sunday morning, as I was getting ready for church, words came to me as if someone were speaking to me."

*Why are you making it so complicated? It's about the people.*

Immediately, Matthew let go. And let God.

NIX is *about the people.*

"When I get to the end of my career or the end of my life, what will make me the most proud won't be the projects we've built or the fancy equipment and big buildings we've added; it will be the team we have nurtured and the lives we have impacted along the way."

• • •

## THE THREE TRUTHS

Another huge breakthrough occurred at the annual strategic planning meeting soon after Matthew realized that the NIX vision boiled down to *the people.*

The planning starts when Matthew and Adam Schmitt, VP of business development and administration, spend one day off-site, thinking long term. "This is our day with our heads in the clouds to really get 'out there,'" Matthew said. "We start the day by reviewing the company vision and deciding if it still makes sense. If the answer is no or 'I'm not sure,' then anything we do throughout the year would be for naught. However, to this day, the answer has always been 'Yes, our vision makes sense.'"

In 2019, Matthew proposed to Schmitt that they spend the entire day answering only one question: "When we get to the end of our career and look back, what would have to be true in order to have achieved our vision?"

When brainstorming, it is valuable to pretend a future state has already been achieved and then look at the roadblocks to be overcome or the successes to be accomplished in order to realize the potential triumph.

Matthew and Schmitt started a list of "truths" and after a couple of hours divided them into buckets to narrow them down. They spent a few hours fiercely debating the options and ended up with three truths that must be established to achieve their vision: culture, growth, and longevity. To achieve the company vision, they would have to pay constant attention to maintaining their corporate *culture*, *grow* the company beyond where it was at that point, and be profitable and operationally *sustainable for the long term*.

Nix Companies' big, hairy, audacious goals and three-year strategic goals all align with the three truths. While at any given time the focus may narrow to one, the team can never neglect any of the three. The three truths are interconnected forces.

Matthew's role became crystal clear. "It is my job as president and CEO to help make decisions that move us toward our vision while steadfastly helping every NIX team member connect their work to the vision, our North Star."

In 2023, CFO Jared Baehl said, "We talk about our big, hairy, audacious goals and the goals that we set as the executive team for 2050, and I think one day, after chipping away for years, we will look up and think, *Wow, this is pretty awesome—what we've been a part of and the team that we built.*"

• • •

## MISSION POSSIBLE

The Nix (parent company) mission reads:

> Be Big, Act Small. To leverage our businesses as a force for good and create value for all stakeholders.

The mission of Nix Industrial (legacy operating company) reads:

> To help keep America's industries safe, efficient, and reliable. We will be differentiated by our culture and strategy of "Be Big, Act Small."

The vision is where the company is headed, and the mission is how it will get there. During onboarding, Matthew tells every new team member, "The Nix mission is how we go to work every day, every week, and every month to achieve our vision *someday*."

The first part of the mission statement refers to the focus of the company's core business, Nix Industrial, which provides safety, efficiency, and reliability through each of its divisions. Some examples are as follows:

- Custom Fabrication builds safety railings and guarding projects for factories.
- Structural Fabrication helps hospitals expand and meet their patients' needs in the safest, most efficient and reliable manner.
- Repairs & Maintenance keeps farm equipment in the fields and tractor trailers on the road.
- Shop Coatings and Field Coatings refurbish and protect the integrity of key equipment for every customer segment.
- Machining repairs and maintains components that keep power plants and heavy industries running and the lights on.

Matthew believes that calling out the company's core purpose—its mission statement—in writing is valuable. He explained:

> First, it reminds all of us exactly what we are doing each day. It's easy to get lost in the hundreds of emails, the miles of welds, the darkness of the sandblast booth, or mired in the frustration of a repair job that's not going as planned. From time to time, it's important to take a step back and see the big picture of what it is we are actually doing. We are doing work that matters.
>
> Secondly, it helps us attract the right kinds of people. If you aren't passionate about our core purpose, then working at NIX is probably not going to be a fulfilling career choice for you. I want people to take a Sunday drive, chest swelled with pride, to show their kids a project they helped finish. If it's just another rail to weld or paint for you, then I don't want our customers' lives depending on you. I want you to connect with what you are doing. Our customers deserve it, and in order for you to have a fulfilling career, you need it.
>
> Lastly, it helps us to always stay focused and in alignment with our core business. There's a saying that more companies die from indigestion than starvation. Now, as I'm saying that, you might think, *NIX does a lot of different things*. Sure we do, but we are very conscientious about who we do it for and whether it fits within our core competency.

The following is a story Dick Farmer, founder of Cintas, shares in his book *Rags to Riches*:

> There was a man walking down the street in the middle of the big city, and he came upon a construction site. Bulldozers and earthmoving machines were busy on the site. People were working hard. He came across three men in a ditch. He asked

the first man, "What are you doing?" "I'm digging a ditch," the first man said. Our protagonist asked the second man, "What are you doing?" "We're digging a ditch for a water line for that building going up over there," the second man said. He asked the third man, "What are you doing?" The man looked up and replied, "We're building a cathedral. It will be a big, beautiful cathedral with five big, tall spires and beautiful stained glass windows. It will seat 500 people. It will be the most beautiful church in this city. That's what we are doing."

Matthew often shares this story with the Nix team and asks the same question Farmer did. "Which of those people do you want on your team? They are all digging a ditch, but one is really proud of and believes in what he is doing. This story demonstrates why it is important for leaders to have a clear, compelling vision and to share it with everyone."

The inclusion of the word "America" in the Nix Industrial mission statement defines the business as nationwide and brings a patriotic dimension. "We are passionate about helping to keep American manufacturing competitive," Matthew explained. "Domestic manufacturing is increasingly coming under pressure from foreign competition. The work we do helps our US customers compete with their foreign adversaries, and we should never understate the significance of that."

"Be Big, Act Small" is the most important phrase for Nix Companies and is much more than a slogan. It's a philosophy that governs nearly every decision. The words resonate with both internal and external stakeholders, team members and customers. As the mission statement suggests, it is both a culture and a strategy.

For a team member, "Be Big" means that through continued profitable growth, NIX provides advancement opportunities, quality of place, and more. "Act Small" means as NIX grows, it will not lose touch with its humble beginnings and will always remember that people are the most valuable resource. Employees treat one another like family. If everyone lives by the core values and protects the company culture, the rest will take care of itself.

For a customer, "Be Big" means NIX will continue to advance their products and service offerings, as well as their quality and safety record, through innovation and continuous improvement. "Act Small" means NIX is nimble, responsive, and honest. As the company grows, it will value each customer, responding promptly to their needs and striving to do business honestly, in order to build long-term relationships.

"Be Big, Act Small." Those four simple words coined by Adam Nix create a philosophical foundation as strong as the iron that founder Charles Henry Nix hammered into shape, forged for a future he could not have imagined.

• • •

## LINDSEY TELLS THE NIX STORY

Jared Baehl became CFO, and after six years as director of finance, Lindsey was finally able to concentrate on sales, developing and servicing the local manufacturing market in places such as Toyota Motor Manufacturing Indiana in Gibson County, Berry Plastics in Evansville, and AstraZeneca, the pharmaceutical company, in Mt. Vernon. She was good at sales and customer relations but did not know the technical side of the business. "I had to drag an engineer with me every time I called on someone," she said.[124]

"I was trying to work four days a week and have one day with our sons, but my phone would blow up on Fridays," Lindsey recalled. "When you are in sales, it never stops. Customers are working on Fridays; they still need us. The sales team kept growing, and I thought, *Our sons are only going to be little for so long*, so I said I had to make a change."

Matthew knew Lindsey would be good at training the sales team and helping the sales reps share the NIX narrative. Lindsey's forte is big-picture customer relations. She is well versed in the history, growth, and future of Nix Industrial. Because she lived much of it.

Today, early on in the sales process, Lindsey accompanies NIX sales representatives on customer calls. "When they have a good initial meeting with a customer, I am the next representative there to help tell the story, answer questions that the sales rep may not be super familiar with. I love to go in and tell the NIX story, help open the door, and then walk away. The sales rep then follows through, supplying the answers to the technical questions, wrapping up the sale, and managing the day-to-day."

Lindsey is outgoing and enjoys her customer-facing role. While maintaining key relationships with the customer base, Lindsey also represents the company and the Nix family in various community organizations and activities and oversees charitable giving. She serves on the Posey County Alliance committee of the Southwest Indiana Chamber, the Junior Achievement board, the Deaconess Foundation board, Catholic Charities, and chairs the Family Business Alliance. "I represent us at those events," Lindsey said. "It used to be Matthew, but we just can't be everywhere." Lindsey and Matthew are also founding members of the Evansville chapter of Legatus, a Catholic business-executive group for couples.

As Matthew's partner in life and business, Lindsey's role goes beyond what is written in her job description, so she well knows how important family buy-in is. "Before we make an acquisition, I want to go with Matthew to meet the owner and their spouse," Lindsey said. "You can tell a lot about a person by their spouse. It's important to us to meet the significant other, particularly in key positions within the company. Getting to know their home life as much as we can in an interview. Because if they don't have support at home, if they are miserable at home, they are going to be miserable at work."

## CHAPTER 27

# Company Culture: The Vibe Inside

Next to the outstanding caliber of the company's people, the NIX corporate culture is the most important aspect of its success. "Culture is what someone *feels* when they spend a day inside our company," Matthew explained. "It's how a new team member or customer *feels* about *us*. The vibe they get."

Company culture is the energy that hums in the background as a worker welds, an administrator plots strategy, or a customer tours the pristine shop buzzing with quality workmanship and attention to detail.

"Culture is like the proverbial iceberg," Matthew said. "There's the part that everyone can see above the water and the part below the surface that's invisible. Much more is below the surface—our beliefs, traditions, past experiences, and assumptions. Below the surface is the lens with which we view the world. Above the surface are aspects such as how we dress, talk to one another, decorate the office, or what kind of vehicles we drive and, most importantly, what behaviors we reward or reprimand."

Unlike vision and mission, which only exist in the future, culture only exists in the present. "You don't get to *decide* today what your culture is," Matthew said. "You can only *describe* it. What your culture is today was decided months and years ago. You can decide to preserve it or even change it, but you will have to be intentional and patient. Culture is a combination of values, traditions, and a constant rolling average of the morale in the preceding sixty days."

"The Nix family truly cares, and you feel like a member of the family," CFO Jared Baehl said in 2023. "That's a testament to the culture that Adam, Matthew, and Lindsey started building with the growth of the company. You feel the culture. Which makes it a joy to come to work every day."[125]

Fortunately, without knowing what they were doing, the leadership team developed over time a noteworthy corporate culture, and outsiders began to ask Matthew to talk about it. "After I was asked a few times to describe our culture, I realized I'd better decide how I wanted to describe it or others would decide for me," Matthew recalled. "Like I often do on these kinds of projects, I spent time while I was away from the office on weekends or vacations with a pen and a yellow legal pad, scribbling down random thoughts. After a while, I condensed them into two concepts that best describe our culture—'progressive and competitive' and 'people and values based.'"

The NIX team when the Indiana Chamber recognized Nix Industrial with an "Indiana Best Places to Work in Manufacturing" distinction in 2023.

The NIX culture is largely centered around the drive to move forward to succeed. Adam likes to say, "We love to win." On the flip side, Matthew describes this ethos as "hating to lose." Losing stings so bad that a person will do whatever it takes to not let it happen.

Another key component is humility. Matthew said:

To remain open minded and never be satisfied, you must be able to recognize how much opportunity you have to improve. The pursuit of continuous improvement, as we like to call it, is what sets many apart. People who are uncomfortable with constant change and challenging the status quo will be extremely uncomfortable within our organization. I have watched this take place. After some failed attempts at coaching, they typically move on, of their own accord. Sometimes we have to help them move on, but we always try to do it respectfully and gracefully. After all, our way isn't the right way any more than theirs is the wrong way. It's simply "our way," and we shouldn't apologize or make concessions for it. Our culture would not make sense in a lot of organizations, even some that might be outperforming us financially, but for us it works. The best way to kill a culture is to allow people to hang around that don't fit within it.

This does not mean the company thwarts diversity. In fact, that would be counter to the NIX culture. In order to be progressive, not only does NIX *want* people who think differently, it *needs* them. NIX needs people with drive, passion, and humility.

The second aspect of the NIX culture, "people and values based," rises from the legacy of the company: *"We are a family business. Let's treat each other like family."*

"We have deep-rooted values and beliefs," Matthew said. "Let's not compromise those for the sake of other pursuits. If we truly believe that our people are our most important asset, then we must act like it."

• • •

## THE EVOLUTION OF NIX CORE VALUES

When Matthew decided to start expanding the company in 2013, he developed the organization's first set of core values, known then as guiding principles, something he put together because "everyone else was doing it." After a couple of years, he realized they did not speak to the essence of what the company valued. So a survey was conducted among the then thirty employees. Questions included the following:

- What words would you use to best describe our company?
- What words would our competitors use to describe us?
- What words would your family or friends use to describe us?
- What words would the community use to describe us?
- What words would you want people to use to describe us?

Matthew spent time over a weeklong vacation with his legal pad, compiling the themes from the input into five core values.

However, he felt something was missing. Then he happened to listen to the podcast series *Entreleadership*. The guest, Dena Dwyer Owens, a CEO of a large home-services franchise, spoke about adherence to core values and how to live them out well. She also talked about behaviors. While core values are important and set a clear standard, their downfall is they are often intangible. A perfect example was the NIX core value of "edge," a general term not defined with any detail an employee could hang their hat on. If they cannot connect to the statement of core values, it is just a collection of nice-sounding, empty words.

Behaviors take the abstract and make it practical. Behaviors are the day-to-day, hour-by-hour actions that support the conceptual ideal. Matthew and the team decided to bring the core values to life so they were not just words hanging on the wall. They incorporated core values and behaviors into many aspects of the company, including evaluations.

Matthew gave an example. "If we were in the evaluation process and I said to John Doe, 'John, you need to work on your *edge*,' he might struggle to connect with what I'm saying. It will likely be counterproductive as he will be frustrated and not able to know how or where to improve. But instead, if I said, 'John, your work ethic is outstanding. You are on time every day. You work overtime when asked, and you never complain. That's edge. However, whenever you are faced with an obstacle, you often stop or get frustrated. Not backing down from a challenge is also edge, so I need you to work on that.'"

"We aren't people who just put fluffy feel-good videos on social media to draw people in," explained Angela Kirlin, director of human resources. "It's real life. It's really how we are. I want people to know our intentions of why we are doing what we are doing."[126]

Angela works every day to do her part to make NIX one of the best places to work, to provide an environment where team members are able to provide for their families, reach their ultimate goals, and go home safe to their loved ones. After a team member has been working at NIX for ninety days, they are invited to the training room, and Matthew and Angela conduct an introduction to Nix history and culture. The team member watches the *NIX History* video, and then Matthew goes into a deep dive on NIX core values, how they came to be, and what they mean.

"Matthew gets it," Angela said. "That's what sets us apart. The president of the company takes the time to do this presentation every single month. He goes into how he prayed about his calling. He explains where our mission and vision come from. Why do we have our core values? How do we use our core values? He tells the team member, 'I hope you retire from here. And I want you to retire making more money than you could have ever possibly imagined. But I hope that is not the number one thing you took from here.'"[127]

The team member completes a survey, "Matthew's Onboarding Questions." The first of six is "Upon retirement, what would have to be true in order for you to say that the money you made here was the least important thing you ever received?" The other queries are as follows:

- What motivates you?
- Describe your perfect "boss."
- Describe your perfect day.

"We take that information, and we really use that," Angela said. "People don't want to be treated like a number. They want to make relationships; they want to form bonds. We plan outside activities to make that happen. The survey helps us build our culture."[128]

There is much below the water in the unseen parts of the NIX iceberg. Outsiders may see the company they once knew as the "welding shop," now spread over Poseyville and the Midwest, and think it looks like helter-skelter growth. However, the below-the-surface mission, vision, core values, and company culture—heavily scrutinized and deliberated—create a deep, wide, and stable foundation for happily employed people and their satisfied customers. The vibe is palpable.

# CHAPTER 28

# Pandemic Shifts

By early 2019, Matthew and the team were on the hunt for the next acquisition. Superior Fabrication was fully integrated into the business, and it was performing well financially. "We were having a string of record years and felt good about taking on our next challenge," Matthew recalled. "Little did we know of the test that lurked around the corner."

Five years earlier, the team had looked at Heritage Custom Fabricators in Princeton, about thirty minutes north of Poseyville, when owner Clint Butts approached Matthew about possibly acquiring the business. Clint's father ran Heritage after buying it from the Woods family, who founded the company. Like NIX, Heritage had been around for a long time and was a staple in the community. At the time, the endeavor seemed like too much of a stretch for Nix, so Matthew politely declined Butts's proposition.

However, by 2019 NIX had grown many times over, while Heritage had stayed relatively the same, making the acquisition more palatable. Heritage was now owned by Sam Przymus, and it was not performing as he had hoped. He wanted to get out from under it. "This would be the first true 'turnaround' that we acquired," Matthew said. "With our strong financial performance and having just experienced acquisition success with Superior, we felt confident to take on this new challenge."

In July, NIX acquired Heritage from Sam, who agreed to stick around as the operations manager. He had a good reputation in the industry and a lot of knowledge, with more than twenty years' experience. He had worked for a large nearby machine shop as the plant manager prior to purchasing Heritage. "We got busy working on a plan to bring this business back to its former glory," Matthew said.

The acquisition was financed, and funds were set aside to replace assets that desperately needed upgrading. In late 2019, a strategy and budget were put together for the next year. "With big plans and high hopes, I announced our grand vision to the Heritage team in early 2020," Matthew said. "Just as we began to put the plan in motion, something happened that no one saw coming. There was a weird virus in China that was killing people. Rumors emerged of the first people in the US to contract the virus. Suddenly it was apparent that the virus was a real problem. By April our world turned upside down with the COVID-19 pandemic. The first thing we did was freeze all of our large capital spending. Until we knew the outcome, we had to batten down the hatches."

Heritage had struggled for years, and hope had been instilled in the team for a brighter future. When the plug was pulled on promised new equipment and improvements, the negative effect of the situation compounded.

"During the lockdown era, we had to keep our management team separated for the safety of everyone's health," Matthew said.

Brandon Wright was the vice president overseeing this operation. "When Covid was in full swing," Brandon recalled, "we made the decision not to move people [between] locations to avoid the risk of spreading the virus. I was reporting through three different sites at the time—Princeton, Rockport, and Poseyville. I had to pick one, so I went to Princeton every day. Then it was easy for me to immerse myself in that business."

To make matters worse, two of Nix Companies' largest businesses, Structural Fabrication (formerly Superior Fabrication) and Custom Fabrication, were both having tough financial years due to the supply chain challenges and material price increases. "We went from the top of the mountain in January 2020, celebrating a record year as a team, to the bottom of the ditch a year later," Matthew said.

Despite being in the midst of the pandemic and the challenges at Heritage, NIX underwent its biggest leap yet. "In late 2020, I came across Northend Gear & Machine, a machine shop for sale just north of Cincinnati in Fairfield, Ohio," Matthew said. "Like Heritage, the company repaired gearboxes and cut gears. However, whereas Heritage focused on larger parts and lower volume, Northend focused on smaller parts and higher volume. We liked the Cincinnati region and surrounding market, where we could cross-sell all of our products and services. This opportunity made a lot of sense, despite the challenges we were facing with Heritage."

Yet NIX did not pursue the opportunity immediately. Roughly a year later, in the summer of 2021, Matthew contacted the broker to inquire if Northend Gear & Machine was still available. "Fortunately for us, they just had another offer fall through," Matthew said. "So we picked back up with our pursuit. The business was operated by the original founders we affectionately refer to as the 'Three Ds'—Duane Ratcliff, Dan Rockenfelder, and Dave Shope."

"The Nix team visited Northend Gear twice," Dave recalled. "The team asked a lot of questions about the operation and the equipment. We then discussed their ideas for moving forward with the deal. We were impressed with how personable and professional they were at the same time."[129]

The Three Ds were machinists by trade and left their jobs in 1988 to start Northend Gear & Machine in a small garage. "At the time of the NIX acquisition, we were getting older, and two of us wanted to transition to retirement," Dave said. "The NIX team was top notch during the acquisition process. There were a few bumps along the way, but we all navigated through them."[130]

"After moving twice and undergoing multiple expansions, the Three Ds built a multimillion-dollar business while raising their families," Matthew said. "The fact that they were still partners at the time of the sale was an amazing feat to me. I hold the deepest respect and admiration for these three guys. They truly built the American dream. In fact, in my office is a framed copy of a newspaper article about these three gentlemen titled 'Their Dreams Came True.' They all hold American flags in the story's photo. As we continue to grow and push for new acquisitions, their story and photo reminds me of the incredible opportunity and the tremendous responsibility we have to honor and build upon the legacies these great people have built."

When he heard how Matthew views Northend Gear, Dave said, "I'm flattered by Matthew's words. Due to hard work, dedication, and sticking to our own core values, we did, indeed, achieve the American dream."[131]

Dave stayed on to help run the operations during the transition. When negotiations were underway, he visited the Poseyville headquarters. "We gave him what I call our 'small-town red-carpet treatment,'" Matthew said. "We want our hospitality to make people feel welcome and appreciated. In fact, today we own a historic bed-and-breakfast called the 'the Founder's Haus' on Main Street to accommodate our customers, suppliers, and team members who visit from out of town, since there are no hotels near Poseyville. The Founder's Haus is a Victorian home built by one of the founding families of Poseyville in the mid-1800s. Jared's wife, Heather, is our host, enhancing the personal touch we try to provide to our guests."

That night after Dave's visit, at dinner, he talked about how impressed he was. "While he is a hands-on blue-collar guy and was impressed by the operations," Matthew said, "it was the people and the company culture that he went on and on about. Although he didn't know precisely what it was, and he could not yet articulate it, we now know he was experiencing our mission and guiding philosophy to 'Be Big, Act Small.'"

"We were very impressed with the business units and how NIX team members interacted with Matthew and Brandon Wright," Dave recalled. "Several other NIX team members took time out of their day to join us for lunch. Throughout our existence, Northend Gear had practiced many of the same core values and principles as NIX. We just didn't articulate them as NIX does."[132]

By October 1, 2021, the NIX team had closed their largest and "furthest from home" acquisition at that time. Dave continues to work at Northend Gear. "I feel that I am in the best place possible," Dave said in 2023. "Nix allows me to blossom as a person and hone my machining and mechanical skills while working for a company that really cares for its employees. One of the biggest takeaways from the NIX experience has been their transparency and communication with their employees. I hold them in high regard and am glad they are the ones who purchased this business that we started and grew for many years. They have proven that they want the same thing for Northend Gear that we did—to grow this business so our people can have a great place to work for many years to come."[133]

"When you approach Cincinnati on I-71 from the south," Matthew described. "You reach a large hill just a couple of miles out and peer down toward a beautiful city skyline sitting just beyond the Ohio River. Evansville, essentially an extension of Poseyville at just twenty miles away, also sits on the Ohio River. That river-town similarity is noticeable. I will never forget topping that hill on a Sunday night and approaching the Cincinnati skyline lit up, shimmering over the river, as we entered the city to officially take over Northend Gear & Machine on Monday morning. It was a symbolic milestone as we transitioned from a small local business to a regional presence. I thought, *The small-town boys are in the big city. Here we go!*"

• • •

## UNFORESEEN CONSOLIDATION

Each acquisition is filled with its own singular challenges, and Northend Gear & Machine was no different, with its fair share of ups and downs. At the time of this writing, Nix Companies is more than two years into the acquisition. Matthew summed up the experience:

It's safe to say it has been a good move for us, and we are only beginning to scratch the surface of the opportunities for our business to expand into the Cincinnati regional market. We typically wait a couple of years between our acquisitions because it takes that long to absorb the new arm of our business and allow the dust to settle, to build strong relationships and understand one another. We try to make as few changes to the business as possible.

Our goal is to allow each business to maintain its autonomy and do what it does well. Our home office provides the support and resources they need. Typically, within a couple of years, those who aren't on board with the changes have found new homes, and those who are on board are often given opportunities to take on larger responsibilities. Our culture isn't a fit for everyone. We prepare for that reality and have come to accept it. For those who embrace the acquisition and the *few* changes that are inevitable, it tends to be a great ride.

Also in 2021, it was apparent that NIX would have to make some serious adjustments with Heritage Custom Fabricators in Princeton, so the team began a plan to consolidate the fabrication aspects of the business back to Poseyville. "This was a difficult process, but if we don't make the hard decisions and changes, a bad situation can become the demise of an entire organization."

The situation improved and slightly stabilized, but soon it was obvious that Heritage would not be viable without significant changes. Brandon approached the executive team and the advisory board and said, "We can turn this thing around, but it's going to take a huge investment and a lot of time." They all knew what he was insinuating: *"We can do this if you want to. But we would have to go all in. We need to ask ourselves, 'Is it the right thing to do? Or should we be investing our time, money, and energy elsewhere?'"*

"Punting is never an easy decision, but sometimes it's the right decision," Matthew said. "Fortunately, by that time, we had acquired Northend Gear & Machine. It was performing well and could take on the gear aspect of the work performed at Heritage."

They devised a plan to consolidate the heavy-machine shop work to the NIX flagship facility in Poseyville. Most of the larger machine work supported the core fabrication business, so having it located within that same complex made sense logistically and helped minimize overhead. The team members remaining at Heritage were offered positions at either the Northend Gear or Poseyville locations. Understandably, most declined, moved on to other opportunities, and were wished the best.

All of the Heritage operation transitioned to the Cincinnati and Poseyville locations. Matthew said,

> In the end, although we never purchased it with the intent to consolidate it, and we worked our way through many challenges, none of us on the executive team regret making the Heritage acquisition. I don't believe we could have started a machining operation from scratch for a lesser investment and ended up where we are today. In hindsight, Brandon's temporary move to Heritage during Covid did not end up driving a turnaround, but it did lead to accelerating tough decisions that needed to be made. Dragging those decisions out would not have been advantageous for anyone. Good leadership is knowing that sometimes deals work, and sometimes you have to make them work. Brandon exhibited good leadership.

The move to Poseyville required significant investment in NIX's Frontage Road flagship facility. In the middle of the pandemic, the team pressed forward with major renovations and additions. To accommodate fabrication, which was the first phase of the move from Heritage, NIX completed the largest-scale facility enlargement up to that time—nearly a $1.5-million high-bay expansion to the fabrication shop.

"Business was tough, and we weren't paying bonuses at the time," Matthew recalled. "That made it hard to tell our team we were going forward with a large-scale expansion, but I saw it as a long-term investment, and one that we needed to execute to bring all pieces together and make us stronger in the short term as well."

Fortunately, the entire executive team and the board were supportive, and they persevered. Within two years of that expansion, Custom Fabrication doubled its revenue and profitability. The next phase was to transition their tenant, Integrated Power Systems (IPS), from the Poseyville facility to the Princeton (Heritage) location. The two businesses essentially swapped places. The Poseyville space formerly rented by IPS was renovated to become the new machine shop, and as of July 2023, the entire operations of what was once Heritage Custom Fabricators was fully consolidated in Poseyville and Cincinnati.

• • •

## ADAM NIX HUNGERS FOR NEW CHALLENGES

Although Adam Nix loved his job and was in his comfort zone, by 2020 he struggled with feeling unfulfilled in his position. "I felt I had more to give to the overall company," he said. "I saw several people whom I had helped hire move up to senior positions above me, and I struggled with that. I was technically an executive on paper and the second-largest shareholder in the business, but I didn't feel I belonged there, and I needed to learn more."

Adam hired Dan Robinson, an executive coach, to push him to acquire more knowledge and hold him accountable in order to progress to a senior level. They met weekly for one year. "He coached me like my dad did my entire life, with a lot of tough love."

In April 2021, Adam began to help lead the Custom Fabrication division from a VP level. This business unit had reported through Brandon Wright, but Brandon was spread too thin with his responsibilities at

Princeton, Rockport, and Cincinnati and needed help. "Just as I did with Repair & Maintenance, I fully immersed myself in learning Custom Fabrication," Adam recalled. In November, he began leading the engineering group. At this writing, Adam is enjoying the challenge of leading the Repairs & Maintenance, Custom Fabrication, and Machining business units in Poseyville, as the vice president of operations.

Adam returned to school and earned his MBA, meeting his goal of achieving straight As. "Balancing running a growing business, raising two daughters, and being a good husband while also working on a postgraduate degree isn't for the faint of heart," Adam admitted. "My faith has helped me considerably, and my wife has been incredibly supportive. I am hopeful I'll look back on these times with fond memories and that my daughters will grow up to see how hard I worked and understand anything is possible if you're willing to put in the effort."

• • •

## ADOPTED TEAM MEMBERS

In 2020, as Matthew and the team brought on new acquisitions, integrating the new "adopted" employees into the NIX culture, he and Lindsey adopted an infant son they named Roman Tyler. Donna, Matthew's mother, said, "Matthew wanted to adopt because he felt he was so fortunate to be adopted that he wanted to give somebody the chance that he had."

Matthew and Lindsey named their son Roman Tyler because they love Rome, Italy, where they had traveled with Father Tyler Tenbarge, Lindsey's cousin, with whom they are close. Matthew's middle name is also Tyler.

"Because we have a family-centric corporate culture and one of our guiding principles is to 'treat team members the way we want our family members to be treated,' plus the adoption of children into our family is near to our hearts, it is fitting that we refer to team members who have joined us through an acquisition as 'adopted'—a warmer and more fitting term than 'acquired team member,'" Matthew said.

Within a ten-year span, Nix Companies Inc. (NCI) fostered six acquisitions and five start-ups with plans for many more to come. They have adopted many new team members into the NCI fold.

## ACQUISITIONS AND START-UPS, 2013 TO 2024:

2013—Industrial Paint & Powder Coating start-up

2015—Tri-County Equipment acquisition

2016—Industrial Field Coatings start-up

2018—Superior Fabrication acquisition

2019—Heritage Custom Fabricators acquisition

2020—Captivated Content (partnership) start-up

2021—Northend Gear & Machine acquisition

2022—Indiana Body Works acquisition

2023—ProFab Alliance start-up

2023—Superior Property Holdings start-up

2024—Kings Machine acquisition

2024—Huncilman Sheet Metal acquisition

## CHAPTER 29

# Irons in the Fire

From 2010 to 2024, Nix Companies grew from four family members running the business to over two hundred team members. It went from one 4,800-square-foot building to multiple facilities totaling over 300,000 square feet. It developed from a local company with 99 percent of its customers in a sixty-mile radius to a regional company with more than 50 percent of its business outside that radius, including customers nationwide.

"We went from a peon small fish that other fabricators and contractors dismissed to a formidable opponent, landing many of the highest-profile projects in the region," Matthew said. "We have gone from an unrecognizable name to a well-respected regional brand known for its fast growth and strong corporate culture. But we are not about to rest on our laurels. We are just getting started."

In 2021, Nix Companies was named the Southwest Indiana Community Champion of the Year by the Evansville Regional Economic Partnership. Matthew, Lindsey, and Adam are passionate about adding value to their community and believe business can and should be a force for good. They serve on many nonprofit boards and contribute their time, talents, and treasures toward these missions, as mentioned previously.

Building on NIX's previous successes and recognitions, in 2022 and 2023, the Indiana Chamber recognized NIX with an Indiana Best Places to Work in Manufacturing distinction. Nix Industrial (in 2024, Nix Metals and Nix Coatings merged into one brand, Nix Industrial) was selected as number thirty-seven on *The Fabricator*'s 2024 FAB 40 list. Established in 1970, *The Fabricator* is North America's leading magazine for the metal forming and fabricating industry. The FAB 40 list is created through data provided by metal fabricators willing to share revenue numbers and company information.

As of 2024, NIX has seven different locations and twelve different business units—acquisitions or start-ups, operating independently as part of the NIX parent company, with ten to sixty people at each. Essentially, it is a series of small businesses that, under the NIX umbrella, further the company mission to "Be Big, Act Small."

Today the three-year strategic plan has many measurable goals, such as increasing ethnic diversity and hiring more women within the operations functions of the business. "We have had some early success and are optimistic about the future," Matthew said. "We have invested heavily into our corporate culture and our recruiting efforts, and I believe we can lead the way in an industry that is traditionally not diverse. Women and African Americans [as an example] do not think of metal fabrication as a common career path. We are looking to change that."

Today Matthew and the executive team are focused on "scaling" the business rather than "growing" it. "There are a couple of differences between growth and scaling," Matthew explained. "First, growth is something that happens to you. It's often less thought out and structured. Typically organic. The second difference is there's a cap on growth, where scaling is somewhat limitless. The only limit is the market cap of the industry. Scaling is more strategic. Planning our future moved us from growth to scaling."[134]

NIX accomplishes scaling with a three-prong strategy:

- Incremental organic growth. Examples: expanding the sales team, improving the company website, improving manufacturing efficiency, etc.

- Transformational organic growth. Examples: launching an entirely new product or service or investing in a new technology, a focus that usually spans multiple years.
- Acquired growth. In other words, acquiring other businesses that fit strategically within the Nix Companies portfolio.

"We are always trying to grow proportionately in each of those three buckets," Matthew said. The foundation that enabled scaling was first laid when Matthew and Adam approached the senior leadership about buying out J. Z. Morris, who had invested in the Industrial Paint & Powder Coating unit. "The company bought him out, took back the stock, and then, over a period of a few years, we sold that stock to our four executive team members," Matthew explained in 2023. "That has been a key component to continuing to scale the organization. We run the business in a more mature manner because we have other owners.

"We all have strengths and weaknesses," Matthew continued. "I'm the visionary, looking for the next opportunity for growth to move the needle forward. We also need integrators. We have tactical people, those good with systems, processes, and controls. The six equity owners in the company are locked in for the long term. We're all rowing in the same direction."[135]

Matthew admits that this equity plan may not be for all companies. "We have just introduced a new partnership model for the next level on the organizational chart because we've seen the fruits of what it's like to have 'owners,' not 'renters,'" Matthew said.

"I like to say, 'How many times have you ever taken a rental car through the car wash?' How do we get more people to feel and act like owners? The challenge is that without enough scale, the ownership interest could get diluted to the point that it no longer achieves the desired risk/reward motivation that it was meant to accomplish in the first place. The only way to mitigate this is to continue to make the pie bigger."[136]

• • •

## NEW ENTERPRISE OFFERS THE NIX "PLAYBOOK"

There are a number of systems, processes, and best practices that need to be in place for a company to scale. Those methods are now part of the NIX core competencies. Matthew explained,

> A couple of years ago, it became apparent that we could help many more shops and owners beyond those we reach through acquisition. We have a playbook that has been developed through trial and error and costly mistakes. That playbook—a collection of philosophies, systems, and processes—became ProFab Alliance, a business venture where Nix Companies leverages its expertise to help others in the metal-services industry continue to grow and succeed.
>
> For example, we spent years struggling with role clarity and division of responsibilities within our businesses until we developed a simple and effective document called "Roles, Responsibilities, & Expectations," or RR&E for short. Today this is a standard template for most of our roles, and we can plug this into any new business or location and add value immediately. We have developed dozens of tools like this along the way that can help others avoid spending months or years making painful mistakes.

Some tools were developed to accommodate the businesses Nix Companies started or acquired through their decentralized model, where each location operates somewhat independently as its own business. "I can't be at every location, but it still needs to run the way we want it to," Matthew said. "Even when owners only have one location, they strive to answer the question 'How can I get my business to perform without me there every day?' The answer leads to a better lifestyle, and if the owner sells the business, it will likely go for a higher value than when it is dependent upon the owner for its day-to-day operation."

The concept for ProFab came in a light bulb moment when Adam

Schmitt was in Matthew's office for his annual evaluation and the two men were discussing the future. Matthew looked at Schmitt and asked, "What is your dream job?"

"To run my own consulting business," Schmitt quickly replied.

His comment was in the context of a "later in life" semiretirement gig. However, Matthew thought about it for a few weeks, then called Schmitt back into his office and said, "Why not do it now?"

The seed of ProFab was born as its own subsidiary business, with Schmitt as managing partner. "It might seem odd that we want to help others in our industry—technically 'competitors.' But we don't see it that way. There's plenty of work to go around for everyone, and we are striving to keep work in the US. I'd rather help my US competitors become stronger. I believe 'a rising tide lifts all ships.'"

It was not all smooth sailing to launch the dream. "We pitched the idea to the rest of our executive team, and they were all in from the beginning," Matthew recalled. "Next we had to convince our board, and that was not as easy. They shut us down on the first pass. We went back to the drawing board, improved our business plan, pitched it again, and got approval to invest a modest amount into the business over each of the next two years."

ProFab Alliance officially launched in 2023. "When our revenue grew a hundredfold, it was the slingshot for ProFab to be created," said Schmitt, who also remains VP of business development and administration for NIX. "We assist with culture, knowledge, and systems to set people up for growth and allow them to take back their life. Our tool kits and resources also help leaders attract and retain the right people who can help a small shop grow."

ProFab Alliance was created with a clear mission: to help other owner-operators within the metal fabrication and machining industries grow their businesses, maximize their profits, and live a more balanced life. Although ProFab shares common ownership with NIX, the operations are run independently, with an intense focus on confidentiality.

ProFab offers three primary solutions:

- The first, consulting and coaching, offers strategic planning with executive coaching opportunities and access to the fabrication-specific management tool kits NIX painstakingly developed. Additionally, it offers fractional leadership and/or back-office support. Instead of taking on the cost to hire a financial professional like a controller/CFO or a head of HR, a client can share a seasoned industry expert with a few others through ProFab's fractional staff services.
- The second is a community network (similar to a cooperative model) where members can participate in industry peer groups and take advantage of shared purchasing benefits, work sharing, and other perks.
- The third is a full franchise opportunity, where a business can be "converted" to the NIX operating system and operate under their umbrella, maintaining independent ownership and primary brand identity with a "Powered by NIX" cobrand. This allows business owners to focus on what they do best—making or repairing parts—while NIX supports by doing what they do best: administration, talent acquisition, retention, and business development. "We want them to go home at the end of the day and have dinner with their family," Schmitt said. "We want people to live the lives they want to live." There's a phrase in franchising that pretty well sums it up: "You are in business for yourself but not by yourself."[137]

• • •

## NIX CONTINUES TO TRANSFORM

In 2023, advanced machining capabilities were added to the flagship Poseyville location, and the following year, the largest expansion in the company's history was underway to make a new home for the Shop Coatings operation. These additions make the flagship Frontage Road

Poseyville operation the ultimate one-stop shop, offering engineering, fabrication, machining, coatings, state-of-the-art inspection, and assembly.

The facility spreads across twelve acres with more than 70,000 square feet under roof, which means more room to accommodate projects of all sizes, and it offers a twenty-five-ton crane capacity, with a building design to accommodate fifty tons in the near future. The addition was something Matthew had dreamed about for nearly two decades. When he started working at the shop, NIX just had one old forklift.

When the crane was delivered and placed in the new facility, it was named after the fourth-generation owner, Charles William "Bill" Nix, Matthew's father. Each crane at the Poseyville Custom Fabrication facility is named after an owner from previous generations. The idea to name all the cranes after the forefathers of the business came from the team members, not from the Nix family. This speaks volumes to the value placed on the company's family roots and longevity.

In summer 2024, NIX acquired its seventh and largest acquisition to date. Ironically another fifth-generation family business, Huncilman Sheet Metal, located in New Albany, Indiana, just across the river from Louisville, Kentucky. This acquisition added more than fifty "adopted team members" to the NIX roster, one of which is Parker Huncilman, the company's fifth-generation family team member who is in his midthirties. Parker joined Nix Companies as the general manager of this business and will retain equity ownership.

"I could not be more thrilled to join forces with the Huncilmans," Matthew said. "Obviously, the idea of joining two fifth-generation metal fabrication families is special to me. I would be lying if I said that was not an emotional appeal, but that is not why we made this deal. This business [at least on paper] may be the best strategic fit of any deal we have experienced. It offers laser cutting, CNC forming, and robotic welding capabilities that will help to round out our metal services portfolio, and its location fits perfectly into our operating geography as it sits almost halfway between Poseyville and Fairfield,

Ohio, near Northend Gear & Machine. It also offers e-coat painting, which marries up nicely with our powder coating capabilities."

"I'm able to progress the company at a better pace than I was capable of on my own," Parker Huncilman said. "So [the acquisition] provided me with a way to limit my risk, excel the company, stick with my employees, invest in them, and become a part of the next brand over the coming years."

"The most critical component of the whole thing is corporate culture and continuity of the team," Matthew said. "The team is everything. We can all go buy machines and buy buildings. But it's about keeping the team intact. Having Parker join us in ownership is really a key element of that in my mind."

On the day of the acquisition announcement, Matthew listened as Gordon "Gordie" Huncilman addressed his team, many of whom had been with the company more than thirty years. Reflecting on the emotion of the event, Matthew said:

> It was the first time I had been a part of such an announcement. Gordie got a little choked up when he was speaking to his team. Afterward he told me that the announcement was the hardest thing he had ever done, next to giving his daughter away in marriage. I said, "Well, you just gave away another baby. I understand."
>
> It really hit home for me how special of an opportunity it is and what a sense of responsibility it is to steward the legacy of the Huncilman family and take care of these adopted team members who have helped build the company into what it is today. I recognize that five generations have poured their heart and soul into this place. Not just the owners, but the entire family. It becomes part of the dinner table. Unless you are in a family business yourself, you just can't fully appreciate that.

Also in 2024, NIX opened a sales and engineering office in Fort

Wayne, Indiana, with two salespeople and one engineer based out of that market. Consistent with its "Be Big, Act Small" philosophy, NIX wants to have a physical presence within a two-hour drive of every major customer. "Having a team in Fort Wayne allows us to effectively serve our growing client base in Southern Michigan, Northern Indiana, and Northern Ohio," Matthew said. "The plan is to add a brick-and-mortar operation there in the near future."

• • •

## HELPING EMPLOYEES WITH CAREER DREAMS

Infrastructure is not the only impressive asset that has progressed during the past few years at NIX. In 2023, NIX also started the Path program to develop career tracks for employees and the Dream Achiever program, which pairs team members with a professional coach to help them develop a list of goals and dreams while providing support and encouragement toward the accomplishment of those dreams. A large "dream board" will be displayed in the new headquarters, listing team members' realized dreams.

"We create a training program for an employee who wants to move up or go into another area," Angela Kirlin, director of HR, said. "It is implemented alongside an employee's evaluation after one year of employment. We're backing it up with tuition reimbursement or commitment to internal training opportunities. It's all about asking, 'Hey, what do you want to do in life?' You might be a welder, but you might aspire to be an accountant. We create a road map for them to get there."[138]

The program builds from existing NIX training programs—one for those aspiring to become company leaders and another for those who focus on the craft of their respective trades. The Path program utilizes large training boards displayed in the shops with the acronym CRAFT: "cultivating real advancement for tradespeople." Each board displays matrices involving people, machines, processes, and technologies that operations managers update each month. Red signifies the person is a novice at a particular task; yellow, an apprentice; green, a master; blue, a trainer. Manuals explain the time, effort, and experience required to move from red to yellow, green, and blue.

"The idea is to have everybody in each shop trained on every single piece of equipment, every single process, and we are cross-training within divisions," Angela said. "Our folks love it. They feel like they aren't stuck on one piece of equipment or one type of project. The goal is that they are diversified in their training and skills. If a person calls in sick, we can pull in someone who has been trained in their area to fill in."[139]

One successful Path employee is Jack Henry Hagan. He started as a welder right out of school in 2018, and five years later, at the age of twenty-five, he was promoted to full-time quality control at the NIX structural shop, responsible for millions of dollars of structural fabrication. At the time of this writing, plans are in place for him to obtain additional training to achieve his certified welding inspector license. Like a proud papa, Matthew said, "There are many more of these stories yet to be told. This is my life's mission."[140]

After his promotion, Jack attended his first quarterly frontline-leaders meeting, where Matthew and Angela meet with all lead people. "When Jack was introduced, Matthew got choked up," Angela said. "He said, 'This is what fires me up every day. To see a team member start on the floor as a welder, move into a fabricator position, then move into a quality control lead.' I share that feeling one hundred percent. We both say it's not about how much money the company can make or the great facilities and great machines; it's about the lives that we are touching. Period. That's the vision."[141]

NIX offers summer internships to students, with opportunities for full-time positions upon graduation and a higher-education tuition-reimbursement program through the Southern Indiana Career Technical Center, local high schools, and many other similar institutions throughout the regions where NIX operates. Internships may be in welding, repairs and maintenance, custom fabrication, structural fabrication, machining, business administration, and engineering. The company also has connections with community and technical colleges.

In 2023, NIX made a full-time offer for an engineering position before the candidate even started college, and he accepted. He received tuition reimbursement and continued to earn an income while working at NIX during his college breaks. As NIX continues to grow, adding more and more people, the concern is that the family feel may be lost. However, Matthew has put initiatives into practice to connect with his people, thanks to one question posed by a board member.

## CHAPTER 30

# Keeping the Family Feel

"'What are your touchpoints?' This was the question one of my board members asked me when I was struggling with the business growing beyond my ability to 'manage by walking around,'" Matthew said.

Because of that one question, I have developed a series of touchpoints, my routines for who I check in with and when and what information I want to know. I review my daily and weekly dashboards. (Those dashboards are now part of what we help our ProFab clients develop for themselves.) However, one of the most meaningful things I do is to host listening sessions.

Once a month, I meet with one small team, five to fifteen members, within our company without the presence of anyone from the layers of the organizational chart between us. I cover everyone in the company every year. Thanks to the mentorship of Jason Lippert, CEO of Lippert Components, which grew from a family business to a multibillion-dollar global company that is now publicly traded—and hearing him speak at the Purpose Summit 2022—I revised my list of questions to include the following:

- What's going well?
- What could be going better?

- What should we start doing?
- What should we stop doing?
- What should we keep doing?

Matthew starts each meeting by asking team members to share some personal big wins. "This is what gets me fired up to keep grinding every day and chase our vision," he said. He hears answers such as "I paid off my credit card," "I lost fifty pounds in the company weight-loss challenge," "I bought my first house," "I bought a new ATV," and "I got engaged."[142]

Jeff Hite, Matthew's brother-in-law and former Cintas sales manager, observes that Matthew is very good at paying attention. "He listens intently before he talks," Jeff said. "Sometimes he doesn't even talk, but he takes a lot in. He's learned that you don't grow big without some type of culture about the people. He's also very good about making family members feel appreciated, valued, and important. That's part of his nature. It's easy for him to genuinely put his people first."[143]

He explained that Matthew is an anomaly. "He's the kind of guy, and they are rare, who wants to create the thing and find all the people who will be a part of what he is trying to create. Matthew is the CEO who created the company, not the CEO who was hired after the company was already successful."[144]

The first annual President's Leadership Forum, where Matthew and Adam, along with others from the executive team, talk about the history of the company and plans for the future, began in 2023. The current and aspiring leaders in the organization have an opportunity to ask questions of one another and the executive team to deepen relationships and create more "touchpoints."

Adam recalled the first forum. "A keynote speaker was needed to talk to our future leaders. We usually bring in outside speakers, and Matthew was trying to think of a speaker that could deliver the right message. After a while, he approached me and said, 'I think you would be perfect for this one.' I made a PowerPoint presentation and gave a talk on what I thought would resonate: how to accomplish goals through grit, finding a way to win, not being afraid to ask for help, getting outside of one's comfort zone, setting goals, and continuing to learn while ultimately enjoying the ride. I told my life story—how, growing up, I never wanted to work in the shop, yet I have fond memories of working with my dad, brother, and aunt."

As a principal owner with Matthew, Adam now looks back at working in the welding shop with some nostalgia. "I still love working with my hands," he said. "If somebody said, 'Hey, Adam, you need to put your old blues back on, go to the shop, and work on a project,' I'd say, 'You got it.'"[145]

The metalworking industry is dominated by family businesses. To say that Matthew and Adam are busy men is an understatement, yet they are in tune with work–life harmony. "There are lots of dads who work a lot of hours and miss stuff with their kids," Lindsey said. "Matthew misses very little. He spends tons of quality time with our kids and does a fair share around the house. Not only is he a great businessman, he is a great husband and a great father. Work–life balance is good. He wants his team to have that balance, and he wants to help other business owners find it too."

Matthew likes the term "work–life integration."

"Perfect balance is like a unicorn," he said. "Especially if you are passionate and on a mission. If I want to take my kids to school or go to Mass on a weekday and go into work late, I do it. If I want to read a business book and brainstorm my next idea while on a family vacation, I do it. In both cases, I hope I am a better leader and father because of it. In either case, I don't pursue some arbitrary idea of balance. When I feel burnt out, I shut work off when I can. When I'm on fire, I fan the flames."[146]

Adam, who volunteers as a head coach for German Township Youth Baseball, has his philosophy about harmonizing his work and his life with Lacey and their daughters, Harper Jean and Hazel Kate, and son, Baker Carl. "Cell phones and computers are a blessing and a curse," he said. "Sometimes after a long day at work, I have a hard time turning it off. But when I wake up in the morning, I'm positive. I come to work with a feeling of optimism. Working with my extended family is great. I don't think any of that is tough. Working with like-minded people and having the opportunity to influence and be responsible for other people and their livelihood is pretty cool."[147]

Adam Nix and his wife, Lacey, with their children, Harper Jean, Hazel Kate, and Baker Carl.

Lacey is an exercise physiologist with the Heart Hospital at Deaconess Gateway in Newburgh. While she has never worked for NIX, she keeps abreast of the business from a high level, hearing about it from Adam as well as through a semiannual written update prepared by Matthew and an annual shareholders' meeting that all spouses are invited to.

"You hear people talk about work–life balance, but they say they do not talk about their work when they are home," Lindsey said. "That is so unrealistic. This is our life. This is our passion. Of course we are going to talk about it at home. When some people hear I work with my husband, they think I'm crazy. To be an entrepreneur, you have to be a little bit crazy."[148]

Matthew and Lindsey rarely let their challenges at home come into the business. "We've done a reasonably good job of not taking work home," Matthew explained. "But we definitely don't take home to work. Even if the 'silent treatment' is going on at home, we will still speak to one another at work like normal. We keep it professional. But we both know that just because we talk at work doesn't mean when we get home everything is fine," he added with a chuckle.

Matthew and Lindsey are charter members of the Louisville chapter of the Legatus Catholic business-executive network. "We call it Catholic date night," Lindsey joked. "Matthew had been feeling like he was missing something in the faith-life portion of his career as a company leader, so he Googled 'Catholic CEO group,' and Legatus came up."

Started by Tom Monaghan, founder of Domino's Pizza and Ave Maria University and former owner of the Detroit Tigers, Legatus is a group of Catholic couples who lead businesses and meet once a month for Mass and a dinner with a featured speaker. When Matthew and Lindsey joined, they were the youngest members; most of the other couples were their parents' age. They eventually helped found an Evansville chapter in order to be closer to Poseyville.

"Seeing how the members of Legatus live their faith and are still very successful is so inspiring," Lindsey said. "They are running much larger businesses than ours. Through their examples, we see that you can be Catholic, have a successful career, be grounded, serve others, and get to heaven. That's our main goal. Get our kids to heaven and get ourselves to heaven. The business is a side part."[149]

Adam's faith plays into his leadership style. "I like to live my life and lead in the way that Christ would approve. People feel comfortable when I am being vulnerable, humble, and positive. I treat people with tough love when they need it, but I do it in a way that's caring. It makes people feel good, and they have a sense of security. When they are at work, they can be vulnerable. They can be themselves. The best team members are the ones who feel as we do—that they are doing life-changing work. Work that matters."[150]

"Adam is the kind of leader who can tell people what they need to hear, not what they want to hear, and they walk away feeling okay about it," Matthew said. "He gives people tough love when they need it, and they respect him even more for it. That's a special gift to be able to interact with people in that way. That's one of his biggest strengths. He's better at that than I am."

As for Lindsey, as director of public relations and training, she initially set a goal to visit each NIX location every month, which meant traveling around Southern Indiana and to Cincinnati. At this writing, she has scaled back to visiting each location quarterly, and she is missed.

"When Lindsey hasn't come around for a while at a location, I hear about it," Matthew said. "That says a lot about her and the culture we have built."[151]

"They ask, 'Where have you been?' And I love that," Lindsey said. "It is so important to our company culture that we are accessible to our people all the time. They have my phone number and my email address; they can reach out to me at any time."[152]

"And some of them do," Matthew added. "Some of them talk to Lindsey about family topics or other issues that they would never talk to me about."[153]

"Sometimes the guys come to me and tell me their wives are pregnant, and she just took a test the day before," Lindsey said. "They are just so excited that they want to tell someone. What a blessing it is for us to be a part of those life events with our folks. As we've grown, keeping this small family feel is a huge and important part of my job. Maybe it doesn't move the needle on the P&L every month, but it helps with turnover and job satisfaction."[154]

Actually, it does move the needle on the profit and loss statement. Team members share work problems with Lindsey that they might not share with their boss, the HR manager, or with Matthew. "That's super valuable, too," Lindsey said. "Because we're getting insight from the shop floor that we may not otherwise obtain."[155]

Sometimes Lindsey worries that people in or outside the company may see her as "just the boss's wife"; however, they would not think such a thing if they saw her step up and give feedback and ideas in meetings. Matthew has fostered her in her leadership journey. "It's helped for me to put myself out there as part of the leadership team," she said. "It's an amazing team that Matthew, Adam, and I built together. It's one of the reasons why we can make working together successful as a married couple."[156]

When Lindsey wonders whether people think her position comes from nepotism, Matthew reminds her of all she did to build the company during its high-growth years, before they had their large leadership team. He reminds her of the sacrifices *she* made. "I try to help her see where her best and highest role is," Matthew said. "I help her see the things that only Lindsey can do. A lot of that revolves around being with our people, being a community servant, a public figure, and a brand ambassador for our company. With her work history and knowledge that no one else has, no one is better suited to fill those roles."[157]

Adam and Matthew also balance being brothers with being business partners. Outside of work, they toggle successfully between both roles. "Some days it's hard to turn it off and turn it on," Adam said. "Part of it is knowing when to laugh and when to be serious, when it's time to cut up and let our hair down and when it is time to be business partners."

Adam shared his thoughts on the possible misperception that

there is family favoritism. "There can be a stigma that we are a bigger company now and our name is on the side of the building. New team members may assume that I did not have to work to be where I am—nepotism. It may be even more of a challenge in the future, but I think it will be okay. I explain to our team members, 'I've been there, I've done that, and I know what you're dealing with.' Hopefully, my legacy is that I care about people. I care about the work they do, and I enjoy the journey more than anything."[158]

The NIX slogan of "Forging Ahead" was developed by Sarah Hausenour, a talented consultant Matthew hired to help with the company's first professional marketing strategy. She wrote:

> To "Forge Ahead" means to move forward and take the lead, to advance rapidly, progress quickly. It also hints at the earliest beginnings of Nix Companies—forged in the humble blacksmith shop of Charles Henry Nix. The generations that have followed have taken his lead, Forging Ahead to build a better, stronger future for our community, our employees, our partners, and our customers.
>
> The forward arrow in the NIX logo points to the future, to forward movement, progress, and ties directly to the tagline below it. This also serves as a rallying cry for the internal team, inspiring them to blaze new trails and focus on the potential that lies in wait.

"Matthew talks about these big numbers that are part of our road map, our vision for the future," Lindsey said. "I think there are people—not those on our executive team, but others—who think he is crazy or that the numbers seem outrageous. But it is going to happen. He's on that track."[159]

Adam does not see himself as a visionary. He says that is his brother's role. "I have no idea what the next decade will look like," Adam said in 2023. "Matthew is doing a good job of trying to paint that picture.

It's my job to continue helping our team turn the vision into reality."[160]

Paraphrased, the Golden Rule is "Treat people like you want to be treated." Matthew tweaked the phrase as one of Nix Industrial's guiding principles. He explained:

> I changed it to "We treat each other the way we want our kids or family to be treated." I made that distinction because as parents, we love our kids, first and foremost, but there are times when you praise them, times when you discipline them, get them back in line, and that speaks to leadership, in general.
>
> And I don't mean we treat our team members like children. But in a family business where you want to have that personal culture, you need to be able to pull someone aside and let them know when they are out of bounds or when something is not acceptable. You also need to be able to love them when they are having a hard time and not delivering the results that they can and should deliver. They may need some grace at that moment. We try to blend those two leadership reactions.[161]

Lindsey often brings her sons to meetings. "We just celebrated one million safe working hours with no injuries, so I brought them in that morning, and we had confetti and silly string while serving breakfast, and they got to see that," she said. "We are laying the groundwork to show them what this special business is and teach them to use the talents that God gave them. Hopefully it will be with our business, but if not, that's okay, too."[162]

"I want people to know our values," Adam said, "and see that we are running this business successfully, the right way, and have kept our families at the forefront. Money is not what drives me to get up every day. I want people to know how much we care about our history, care about our people, who are part of it with us, and how much more is left to be. My five-year-old daughter, Harper, enjoys coming into my office. She helped me hang shelving and photos on the wall. She loves seeing the NIX logo. We have a company ad on the back of our church bulletin. During Mass she points at it and says, 'NIX—that's Daddy's work!'"[163]

In December 2022, Matthew and Lindsey's oldest son, Charlie, began working in the shop, "cleaning chips" from the floor, just as Matthew, Bill, Sonny, and Carl had done before him. Just like the summer day when Matthew, age eight, entered the welding shop, called in from daycare to help his father with a repair that necessitated small, eager hands.

Matthew and Lindsey Nix with their children, Charles "Charlie" Matthew, Mason Eugene, and Roman Tyler.

The day the sixth generation of the Nix family entered the building and got to work offered a storybook metaphor for the circle of life, for it was the very day the fourth generation of the Nix family was laid to rest: Sonny Nix—little Charlie's great-grandfather and the grandfather whom Matthew and Adam had worked alongside since they were boys.

"Charlie was out of school that day for the funeral," Matthew said.

"After a beautiful ceremony and the bereavement lunch, he asked me if he could go to work with me for the rest of the afternoon. Perhaps he was inspired by the outpouring of love, respect, and admiration he had seen for his great-grandfather. Perhaps, somewhere in his little, intuitive eight-year-old mind, he knew that was just what his father needed. Regardless, it was the perfect tribute to his father's hero and the perfect beginning of the next chapter."

"Charlie walked into the house that afternoon after being in the shop," Lindsey remembered. "And the smell. I thought, *Oh, my gosh! This is what your dad smelled like when he came home after working in the shop.*"[164]

"Cutting fluid and welding smoke. It smells like money," Matthew explained with a laugh.

The generational pull, the nod to the Nix legacy, is profound. Sometimes when the stress begins to compound for the couple, Matthew looks at the big picture, thinks of the future, and draws Lindsey into his dream with these words: "How awesome is it going to be when our boys are grown and see what we and our team have created with this business? Think how proud they can be when they say, 'Man, Mom and Dad were really around a lot, and they still did this big thing.'"[165]

The definition of the word "nix" may be "nothing" or "to stop," as Sonny Nix once told Matthew's mother, Donna. However, today, "NIX" means so much on so many levels.

NIX is a "big thing" because the team *acts small*, placing people first, as it forges into the future.

The Nix Industrial team in 2023.

# EPILOGUE
by Matthew Nix

## TO LEAD IS TO MAKE MISTAKES

Sharing our story in writing is one of the most unnerving things I have ever done. It's uncomfortable for several reasons. First, no one loves to be vulnerable and air out "dirty laundry." I wrestled tremendously with the inclusion of some parts of the story but ultimately agreed with the writing professionals that the narrative would not be authentic if some aspects were omitted. And perhaps those omissions would contain some of the best lessons.

Secondly, I don't want people to get the idea that we are better than we are or for the book to present some sort of pumped-up ethos around me/us. This is particularly uncomfortable when you are from a small town. I had to make peace with the fact that while many in our community will be delighted by our story and thankful for our vulnerability in sharing it, others will roll their eyes and likely be on the lookout for any chance to point out blunders, anything that appears to contradict our beliefs and values so often mentioned in the book.

It's important for me to point out that I have made a lot of mistakes. Some of which were profiled in this book, but many were not. I continue to screw up all the time. The fear that I won't live up to the persona that's been documented here is something I must grapple with and beat back. In the end, I find solace in knowing (at least in my mind) that I *will not* live up to it.

My confidence and will to press forward comes not from an expectation of eliminating mistakes but in knowing that I will encounter each blunder with the values and faith instilled in me from those I was blessed to learn from, as well as the wisdom, counsel, and tenacity of a world-class team beside me.

Lastly, there's a fact that's often taken for granted: Even if we make more good decisions than bad and we manage the company well, many factors are often outside of our control. Not the least of which is our own health. It's not lost on me that our family and our team have been blessed with good health, and as a company we have (thus far) avoided any major incidents beyond our control.

To keep the pace we have kept and raise young families requires good health—mental, physical, and spiritual—and an outstanding support system. That has largely consisted of our parents, all of whom are willing and able to help. This makes a tremendous difference. Take out any of these variables, or insert just one unforeseen circumstance, and our "second half" may not be as story worthy. Then again, maybe it will. Anyway, that is (to some degree) outside of our control. Which is precisely my point.

• • •

## GRANDPA'S FINAL WORDS

In the summer of 2022, we learned that Grandpa Sonny Nix had cancer of the esophagus. He sprang the news on me at our biannual meeting, just minutes before I had to stand in front of our entire company and deliver a speech. I thought he was as healthy as a horse. Adam and I had just played a round of golf with him weeks earlier.

There are only a couple of occasions a year where I can speak to our entire team, and I want them to leave feeling inspired. I had to stand up and give them a pep talk when all I wanted to do was cry. That was a hard day but a perfect metaphor for the struggles of business. There will be days when we get punched in the mouth and have to wipe the blood from our lip, grit our teeth, and forge ahead.

In Grandpa's last days, I brought Fr. Tyler Tenbarge and Fr. Tony Ernst, friends of mine, to visit and give him a blessing. It was painful for Grandpa to talk, and he chose his few words wisely. I heard him faintly tell Fr. Tony, "No remorse, no regrets." Grandpa had no regrets for his life.

On the final visit Adam and I made before Grandpa lost consciousness, we said our goodbyes, not sure if we would see him again. The last words he uttered to us took every ounce of energy he had. We leaned in. Faintly, he said, "Do your thing, boys. And do it well."

I've spent a lot of time thinking about those words. Of all the things Grandpa could have said to us, why that?! And what exactly is "our thing"? After some reflection, I realized that "our thing" is what we are already doing. Chasing our vision. Striving to operate many small-to-medium-size job shops, helping others, and creating value in our world.

Dad, Aunt Caroline, Adam, and I (along with several other family members) were all with Grandpa at his home on his final day on earth. We held his hand until he took his last breath at 2:30 p.m. on December 16, 2022.

We stayed at his home until the hospice nurses prepared his body to be taken to the funeral home. After he was gone, I asked Adam what he wanted to do. "Do you want to go home?" For some reason, neither of us wanted to. So, we did the only thing we knew to do: we went back to work.

It might seem odd that we went back to work less than an hour after our loved one passed. But I suspect those in a family business or anyone who loves what they do will understand. To us, it was the only thing that felt right. When we lose someone, we want to be where we are comfortable, with people we love. So that's where we went.

When we arrived at work, it was nearly quitting time for our shops, and it was Friday, so we had a couple of beers with some of our team as they ended their work week. They had no idea that an hour earlier, we had been holding Grandpa's hand as he took his last breath. We did not bring it up. There was no reason to discuss it. He was home with our Lord, and we were where we belonged. We were with our people. We were doing our thing.

Four generations of the Nix family gather inside the custom fabrication high bay at Frontage Rd, Poseyville, Indiana, two months before Sonny Nix passed away.

• • •

## SHAPING THE FUTURE

As we launch our third strategic plan in 2024, I am excited about our future. We have the strongest team and asset base in our company's history. Even though I just celebrated my twenty-year anniversary working full-time in the business, and it has been more than thirty years since I began to help Dad and Grandpa part-time in the shop, I still learn something new almost every day. Sometimes it is the hard way, but I take comfort in the many mistakes and lessons that lay behind me.

I am nearing my fortieth birthday, taking stock of my life and what I want to do with the "second half." A recent book has inspired me to embrace a new mindset to propel my life and Nix Companies on a course to true significance.

On a cold Sunday afternoon in January 2023, I was relaxing in my reading chair after an epic weekend at French Lick Resort, where we had just celebrated the achievement of our "100X" growth goal with our leadership team and their spouses. I broke open a book that a respected friend had referred to me weeks earlier. I soon discovered that it was no coincidence that I chose that day to start reading *Halftime: Moving from Success to Significance* by Bob P. Buford.

Just like the very first leadership book—*Leading with Your Legacy in Mind*—that I picked up when we were in the hospital with our firstborn, Charlie, the timing was right, and the soil was fertile. The book I was about to read would change my life.

On the first page of the introduction to the book was the subheading "100X." It refers to the scripture found in Mark 4:3–8, which relates the parable of the "sower" who scatters seeds in soils of varying fertility. They say sometimes God whispers to us, but on this day, I was groggy from a big weekend celebration, and He took no chances. He hit me upside the head with a two-by-four. I read that subheading and thought, *Okay, Big Guy, you've got my attention.* However, just as the scripture passage infers, the seeds this book planted did not bear fruit overnight. It took me months of discernment to realize how I was being called to live the message. Soon after, I developed a list of goals for how I would use the gifts God has given me to return everything to him 100 times over.

One of my goals is to help 100 people realize their dream of owning their own business. The franchise business model was something that I had been toying with for a couple of years. The clarity of purpose the book *Halftime* provided me cemented the franchise concept as an endeavor I am called to. Our franchising business won't be the only way that I realize this personal goal of helping 100 people realize their dream of business ownership, but I am confident it will be a big part of it. (For example, I am currently working with our head of marketing and her husband to help them open their own gym and fitness center in our town.) Franchising is going to change dozens, if not hundreds, of lives, and it's going to make waves in our industry.

Moving from a medium-size regional company to a large national company is unnerving, but that can't be reason enough to not go there. The world is starving for larger companies that operate in a principled, people-centric manner. It bothers me to think that we could grow so large that I am unable to know everyone in our organization on a personal level or even know everyone's name. However, I recognize that to achieve our vision and stay on mission, I will have to leave that worry behind. If I want to come to the end of my life and be able to say the words Grandpa could say—"No regrets"—I must forge ahead.

In spring 2023, Lindsey and I, along with Adam and his wife, Lacey, were together on a business trip in New England. As we drove just north of Boston, we saw a Jasper Engines delivery truck. Jasper Engines is headquartered in Jasper, Indiana, a small town less than an hour from Poseyville, and has humble beginnings with similarities to our company culture. Today, they are a nationwide, employee-owned company grossing over $1 billion in annual revenue. Their president, Matt Weinzapfel, now sits on our board, and Jasper Engines was the same size we are today when he started there.

Later that day, we saw a Cintas truck. As you know from reading this book, we have a strong connection to Cintas, and its longest-tenured executive, Greg Eling, sits on our board. Seeing those two trucks was impactful, a nod to what is possible, as we strive to create our national brand.

"That's going to be us," I said. "When our kids are driving down the road someday on vacation in a different part of the country, they are going to see one of our trucks on the road."

It's wild to think about, but I believe it's going to happen. If you doubt us, that's okay. As my dad would say, "Just hide and watch."

# ACKNOWLEDGMENTS
## by Matthew Nix

To my wife, Lindsey, "someday" is already here. It's worth every ounce of effort we have put in. Not only would I not want to do this without you, I *could not* have done this without you. The best sale I ever made was convincing you to say yes.

To our executive team partners—my brother, Adam; Brian Merkley; Adam Schmitt; Brandon Wright; and Jared Baehl—thank you for believing in my vision, executing the mission, and always having my back. We make a hell of a team.

To our entire team, each of you has contributed to this story, and I am so grateful. For those of you who are "OGs", or members of our leadership team who were not called out by name simply because the narrative didn't evolve that way (You know who you are.), you are a monumental part of this history. We have been through a lot together. You are family.

To my parents, thank you for the foundation you built for us. Not only with the business but also, more importantly, in the example of character, commitment, faith, and values you instilled in us at home. Also, thank you for contributing and agreeing to share our story.

To Lindsey's parents, Don and Elaine Tenbarge, thank you for the important work you do behind the scenes to allow us to chase our dreams. We couldn't do it without you.

To those who have trusted us to carry on their business legacy and steward their most precious business resources, their team members, we are deeply grateful, and the sense of responsibility is not lost.

To our outstanding writer, Angie Klink, thank you for your interest in our story and for sharing your time and talents with us. Thank you for the countless edits and for never growing frustrated with me. You were first-rate throughout the entire project.

To Koehler Books, thank you for taking a chance on us.

To all those who have contributed or supported Nix Companies in one way or another—team members, subcontractors, suppliers, those we've adopted through acquisition, community members, or friends who have cheered us along—thank you. I'm literally moved to tears when I think about all the people who have made this story a reality. As they say, "It takes a village," and in tiny Poseyville, we are certainly proof of that.

—Matthew Nix, November 2023

# APPENDIX

## CORE VALUES: HOW WE BROKE THEM DOWN, PLAIN AND SIMPLE

In 2023, the NIX leadership team met in New Harmony for a daylong off-site meeting. One of the topics covered was a revision of the company's core values. The primary motivation for the company's revision was to ensure it was stated in "plain, simple English that would resonate with the average team member." The following is the company's "Core values 2.0":

- Edge
- Doing What We Say and Doing It Professionally
- Caring for Others
- Humble Team Players
- Fun and Fulfilling Work

## EDGE

Edge is where the passion, drive, and grit of an entrepreneur meets the character and extreme ownership of a leader. We are hungry to grow and add value. We are gritty and resilient when faced with challenges. We share a spirit of ownership in our results. We have a whatever-it-takes attitude and a bias toward action but remain disciplined in our approach.

*Behaviors we value*: drive, work ethic, passion, and excitement for what we do, open-mindedness, innovative/outside-the-box thinking, a whatever-it-takes attitude, resourcefulness, nimbleness, responsiveness, not backing down from a challenge, offering solutions instead of problems, decisiveness, discipline.

## DOING WHAT WE SAY AND DOING IT PROFESSIONALLY

We value honesty, integrity, and professionalism. We follow through with commitments. Our yes means yes and our no means no. We treat others with respect and professionalism. We hold ourselves to a higher standard.

*Behaviors we value*: honesty, consistency, follow-through, keeping our commitments, candid but respectful communication, promptness, loyalty, a professional image.

## CARING FOR OTHERS

We treat each other the way we want our family to be treated. We take care of each other and our community.

*Behaviors we value*: compassion/extending grace, thoughtfulness, health and safety awareness, sharing resources, service to others.

## HUMBLE TEAM PLAYERS

We are humble enough to ask for help when we need it. We know we can grow and improve individually and as a group. We are all "one NIX" and know we are better together. We leverage each other's strengths to accomplish our vision. We never say, "That's not my job."

*Behaviors we value*: saying "I don't know," asking for help, offering help, continuing education, making decisions for the greater good of the team.

## FUN AND FULFILLING WORK

We enjoy each other and the journey. We celebrate the wins and don't take ourselves too seriously. We value leaders who keep work fun and help others find meaning and purpose.

*Behaviors we value*: a sense of humor, keeping it real, celebrating wins, appreciating the journey, camaraderie, sharing the passion and love for what we do.

• • •

## CORE VALUES: REINFORCED IN MANY AREAS OF THE COMPANY

*Prescreening and Hiring*: Core values are integrated into the hiring process at all levels. "This is the most difficult part of any interview process," Matthew admitted. "It's relatively easy to develop testing to assess how someone can perform assigned tasks. It's difficult to assess their work ethic, honesty, whether or not they care for their neighbor, their ability to work with a team, or if they are generally fun to be around. Even so, it's still important that we do our best to prescreen based on these attributes—knowing we won't always get it right."

*Evaluations*: During onboarding, employees are made aware that evaluations will consist of two types of feedback: performance regarding actual work tasks and performance regarding the core values and associated behaviors. During an evaluation, they are praised for how they excel and lead others by example but are also made aware of where they need to grow.

*Pruning the Branches*: Unfortunately, even after careful hiring and coaching, not everyone is cut out for the NIX culture or aligned with the values. And that is fine. To be fair to them, other employees, and the organization, they and the company must part ways. Branches must be pruned for the tree to grow full and healthy.

*Team Meetings*: The core values are shared in an authentic manner in meetings. Every Tuesday, fifty-two weeks a year, the entire company comes together for five to ten minutes, with some employees on a video monitor via Zoom. One of the discussion topics is "core value shares."

Each week, employees submit the name of a team member they have observed living out a core value.

For example: Bob stayed late to finish a project. The value is "edge," and the behavior is a whatever-it-takes attitude. The "core value share" builds Bob up while fostering camaraderie and positive reinforcement of the value. All nominators and nominees are entered into a drawing held at a company meeting held every six months. One nominator and one nominee win $100 and recognition.

*Big Decisions*: Big decisions are evaluated through a financial lens and how they align with core values. One example is "Caring for People." If the company chooses not to invest in quality-of-place initiatives or safety protocols because those practices do not contribute to the bottom line, there is a glaring need to reevaluate what is "core" to the company's principles.

• • •

## WORK THAT MATTERS

Today NIX serves a diverse set of customers nationwide, providing critical services and infrastructure that underscores the company's commonly used slogan, "work that matters." With a list like the following, the team can easily correlate their work to a higher purpose. Nix Industrial's "Work That Matters" includes the following:

- Access platforms for backup generators on a data center in Palo Alto, California.
- Structural steel for the first Columbia-class submarine (the newest class of US naval submarines).
- Support structures and components for the manufacturer of nuclear propulsion on US naval submarines and aircraft carriers.
- Piping, plant maintenance, and fabrication of critical infrastructure for Duke Energy, TVA, AEG, Florida Power and

Light, Center Point, and other utilities throughout the Midwest and the South. We literally help keep the lights on.
- A fabricated and powder coated part on top of the Empire State Building.
- Refurbishment of a tugboat for the US Coast Guard.
- A structural steel mine shaft and elevator for a copper mine in Nevada measuring 2,400 ft deep. (If it were aboveground, it would be taller than the Empire State Building.)
- Tank coatings and other critical maintenance for various Midwestern refineries.
- Tank maintenance and other fabrication for Buffalo Trace Distillery and Barton Distillery.
- Plant maintenance and infrastructure as well as manufactured parts for Toyota, Tesla, Ford, and General Motors.
- Gear box rebuilds for some of the nation's most important rotating equipment.
- Short-run manufacturing for various OEMs of products, including agriculture tractors, lawn and garden, forestry, construction, and mining equipment.
- New construction (structural steel) for hospitals and health care throughout the Midwest, including Baptist Health, Norton Health Care, IU Health, Deaconess Healthcare, and the VA.
- New construction (structural steel) for higher education throughout the Midwest, including Indiana University, University of Kentucky, and Southern Indiana University.

## WE WOULD LOVE TO CONNECT WITH YOU

**Learn more about Nix Companies, career opportunities, or the acquisition process.**
From our Humble Beginnings, we are... Forging Business Ahead

NIX — nixcompanies.com

**Work with our coaching & consulting team or discuss opportunities for franchising.**
In Business For Yourself, Not By Yourself

ProFab ALLIANCE — profaballiance.com

**Schedule Matthew Nix as a guest speaker or listen to his podcasts, where many topics from this book are discussed.**

Be Big, Act Small.  @matthew-nix
bebigactsmall.com

**Talk with author Angie Klink or learn about her other books & work as a writer.**

ANGIE KLINK
Author, Historian, Copywriter, Scriptwriter — angieklink.com

# ENDNOTES

1. Todd Bridigum, *How to Weld*, (Minneapolis: Motorbooks, 2008), 10.
2. Ibid.
3. Ibid, 12.
4. Author interview with Bill Nix, April 4, 2023.
5. History provided by Karla Rice, great-granddaughter of Charles Nix.
6. "The Making of a Steinway," Steinway & Sons, accessed May 26, 2023, https://www.steinway.com/news/features/inside-the-new-york-city-factory
7. Karla Rice History.
8. Harold Martin, "Passing of an Era Saddens Blacksmith," *Evansville Press*, December 25, 1977, 1.
9. "Posey County Gives Ovation," *Evansville Courier & Press*, September 11, 1908, 1.
10. Ibid.
11. "Representative Nix Brings Up Sons to Work," *The Evansville Courier*, August 10, 1913, 5.
12. "Biography: Andrew Carnegie," *American Experience*, accessed May 30, 2023, https://www.pbs.org/wgbh/americanexperience/features/carnegie-biography/.
13. "'Pocket' Crumbs," *The Boonville Standard*, December 28, 1906, 2.
14. "Veteran Organ Builder Dies," *Evansville Press*, December 26, 1928.
15. Organ History, St. Francis Xavier Church, accessed May 30, 2023, http://www.evansvilleago.org/organs/in_posyv_francis_xavier.htm.
16. Karla Rice History.
17. Harold Martin, ibid.

18  Ibid.
19  Ibid.
20  Ibid.
21  Ibid.
22  Ibid.
23  Caroline Nix Eickhoff, *A Nix History* Video, accessed June 6, 2023, https://www.nixmetals.com/who-we-are/our-history/.
24  "Nix Welding Moving to New Location," *Evansville Press*, August 18, 1957.
25  Author interview with Bill Nix, April 4, 2023.
26  "Accident Survivor in Improved State," *Evansville Courier & Press*, September 30, 1947, 14.
27  Bill Nix, ibid.
28  Rich Davis, "Memories of Bull Island," *Evansville Courier & Press*, September 6, 1992, 57.
29  Ibid.
30  *Sonny, A Nix Legacy*, Video, January 4, 2023, https://www.facebook.com/nixcompanies/videos/442143551336657.
31  *Sonny, A Nix Legacy*, ibid.
32  Author interview with Bill Nix, April 4, 2023.
33  Sonny Nix, *A Nix History* Video, accessed June 6, 2023, https://www.nixmetals.com/who-we-are/our-history/.
34  Caroline Nix Eickhoff, *A Nix History* Video, accessed June 6, 2023, https://www.nixmetals.com/who-we-are/our-history/.
35  Bill Nix, ibid.
36  Author interview with Donna Nix, May 5, 2023.
37  Sonny Nix, ibid.
38  Eickhoff, ibid.
39  *Sonny, A Nix Legacy*, Video, January 4, 2023, https://www.facebook.com/nixcompanies/videos/442143551336657.

40   Author interview with Bill Nix, April 4, 2023.

41   Ibid.

42   Ibid.

43   All Matthew Nix quotes in this chapter are from an interview with the author on June 16, 2023.

44   Author interview with Donna Nix, May 5, 2023.

45   Author interview with Lindsey Tenbarge Nix, May 15, 2023.

46   Ibid.

47   Ibid.

48   Ibid.

49   Jim Collins, *Good to Great, Why Some Companies Make the Leap …and Others Don't*, (New York: Harper Collins, 2001), 39.

50   Ibid, 37.

51   "Married to Each Other and the Business with Matthew Nix Part 1, *Making Chips* podcast, July 26, 2023, https://www.makingchips.com/listen/married-to-each-other-and-the-business-with-matthew-nix-part-1-369?utm_content=258080904&utm_medium=social&utm_source=linkedin&hss_channel=lcp-3788015.

52   Email from Matthew Nix to Author, June 9, 2023.

53   Author interview with Bill Nix, April 4, 2023.

54   Allison Butler, "Maiden Voyage," *Evansville Living*, April 8, 2010, https://www.evansvilleliving.com/maiden-voyage/

55   Trisha Weber, "Heavy Metal," *Evansville Business*, December/January 2013, 30.

56   Author interview with Lindsey Nix, May 15, 2023.

57   Sara Anne Corrigan, "Posey County Crew Hand-Built Yacht Ready for Sea," *Evansville Courier & Press*, June 13, 2010, 38.

58   Ibid.

59   Lindsey Nix, ibid.

60   Allison Butler, ibid.

61  Email from Matthew Nix to author, June 26, 2023.
62  Author interview with Lindsey Nix, May 15, 2023.
63  Sara Anne Corrigan, "Posey County Crew Hand-Built Yacht Ready for Sea," *Evansville Courier & Press*, June 13, 2010, 38.
64  "Mount Vernon," Ports of Indiana, accessed June 26, 2023, https://www.portsofindiana.com/locations/mount-vernon/.
65  Author interview with Donna Nix, May 5, 2023.
66  Donna Nix, ibid.
67  Lindsey Nix, ibid.
68  Sara Anne Corrigan, ibid.
69  Allison Butler, "Maiden Voyage," *Evansville Living*, April 8, 2010, https://www.evansvilleliving.com/maiden-voyage/.
70  Trisha Weber, "Heavy Metal," *Evansville Business*, December/January 2013, 30.
71  Lindsey Nix, ibid.
72  All quotes from Matthew Nix in this chapter taken from an email from Matthew Nix to author, June 27, 2023.
73  Author interview with Don Tenbarge, May 16, 2023.
74  Author interview with Lindsey Nix, May 15, 2023.
75  Ibid.
76  Ibid.
77  Author interview with Don Tenbarge, May 16, 2023.
78  Author interview with Adam Schmitt May 16, 2023.
79  Denny Simmons, "A Century in Groceries," *Evansville Courier & Press*, December 28, 2015, 4.
80  Schmitt, ibid.
81  Ibid.
82  Ibid.
83  Author interview with Lindsey Nix, May 15, 2023.
84  Schmitt, ibid.

85  "Our Purpose," ADM, accessed July 5, 2023, https://www.adm.com/en-us/about-adm/.
86  Author interview with Brian Merkley, May 16, 2023.
87  Ibid.
88  Ibid.
89  Ibid.
90  Author interview with Lindsey Nix, May 15, 2023.
91  Ibid.
92  Brian Merkley, ibid.
93  Author interview with Don Tenbarge, May 16, 2023.
94  Author interview with Bill Nix, April 4, 2023.
95  "Married to Each Other and the Business with Matthew Nix Part 1, *Making Chips* podcast, July 26, 2023, https://www.makingchips.com/listen/married-to-each-other-and-the-business-with-matthew-nix-part-1-369?utm_content=258080904&utm_medium=social&utm_source=linkedin&hss_channel=lcp-3788015
96  Author interview with Adam Nix, May 15, 2023.
97  Author interview with Brandon Wright, May 15, 2023.
98  Ibid.
99  Author interview with Lindsey Nix, May 15, 2023.
100 Ibid.
101 Jim Collins, *Good to Great, Why Some Companies Make the Leap ... and Others Don't*, (New York: Harper Collins, 2001), 90.
102 Author interview with Jeff Hite, October 25, 2023.
103 Jim Collins, *Good to Great, Why Some Companies Make the Leap ... and Others Don't*, (New York: Harper Collins, 2001), *202*.
104 Author interview with Lindsey Nix, May 15, 2023.
105 "Nix Metals Announces Agreement to Acquire Superior Fabrication, Inc." December 19, 2017, https://www.nixcompanies.com/nix-metals-to-acquire-superior-fabrication-inc/.
106 Lindsey Nix, ibid.

107 Author interview with Angela Kirlin, May 16, 2023.

108 Ibid.

109 "What is Cursillo," Diocese of Evansville Cursillo, accessed August 8, 2023, https://www.evansvillecursillo.com/what-is-cursillo.

110 Ibid.

111 Married to Each Other and the Business with Matthew Nix Part 1, *Making Chips* podcast, July 26, 2023, https://www.makingchips.com/listen/married-to-each-other-and-the-business-with-matthew-nix-part-1-369?utm_content=258080904&utm_medium=social&utm_source=linkedin&hss_channel=lcp-3788015.

112 Ibid.

113 Author interview with Jared Baehl, May 16, 2023.

114 Ibid.

115 Author interview with Lindsey Nix, May 15, 2023.

116 Ibid.

117 Richard T. Farmer, *Rags to Riches: How Corporate Culture Spawned a Great Company*, (Wilmington, Ohio: Orange Frazer Press, 2004), xi.

118 Author interview with Angela Kirlin, May 16, 2023.

119 Lindsey Nix, ibid.

120 Kirlin, ibid.

121 Ibid.

122 Ibid.

123 Ibid.

124 Author interview with Lindsey Nix, May 15, 2023.

125 Author interview with Jared Baehl, May 16, 2023.

126 Author interview with Angela Kirlin, May 16, 2023.

127 Ibid.

128 Ibid.

129 David Shope email to author, November 7, 2023.

130 Ibid.

131 Ibid.

132 Ibid.

133 Ibid.

134 Matthew Nix, "Can You Scale Rapidly after 100 Years of Slow Growth?" Making Chips Podcast, April 11, 2022, https://podcasts.apple.com/us/podcast/making-chips-podcast-for-manufacturing-leaders/id953541032?i=1000557137747 Note: Matthew began as a guest host on Making Chips podcast in 2023.

135 Ibid.

136 Ibid.

137 Ibid.

138 Author interview with Angela Kirlin, May 16, 2023.

139 Ibid.

140 Matthew Nix LinkedIn post, accessed, May 2023.

141 Kirlin, ibid.

142 Ibid.

143 Author interview with Jeff Hite, October 25, 2023.

144 Ibid.

145 Author interview with Adam Nix, May 15, 2023.

146 Matthew Nix LinkedIn post, July 2023.

147 Adam Nix, ibid.

148 Married to Each Other and the Business with Matthew Nix Part 1, *Making Chips* podcast, July 26, 2023, https://www.makingchips.com/listen/married-to-each-other-and-the-business-with-matthew-nix-part-1-369?utm_content=258080904&utm_medium=social&utm_source=linkedin&hss_channel=lcp-3788015

149 Author interview with Lindsey Nix, May 15, 2023.

150 Adam Nix, ibid.

151 Ibid.

152  Ibid.

153  Ibid.

154  Ibid.

155  "Married to Each Other," Part 2, July 31, 2023, Ibid.

156  "Married to Each Other," Part 1, Ibid.

157  Part 2, Ibid.

158  Adam Nix, ibid.

159  Lindsey Nix, ibid.

160  Adam Nix, ibid.

161  Part 2, ibid.

162  Ibid.

163  Adam Nix, ibid.

164  Ibid.

165  Ibid.

Printed in the USA
CPSIA information can be obtained
at www.ICGtesting.com
LVHW090538151124
796532LV00027B/105/J